PLATE. I.

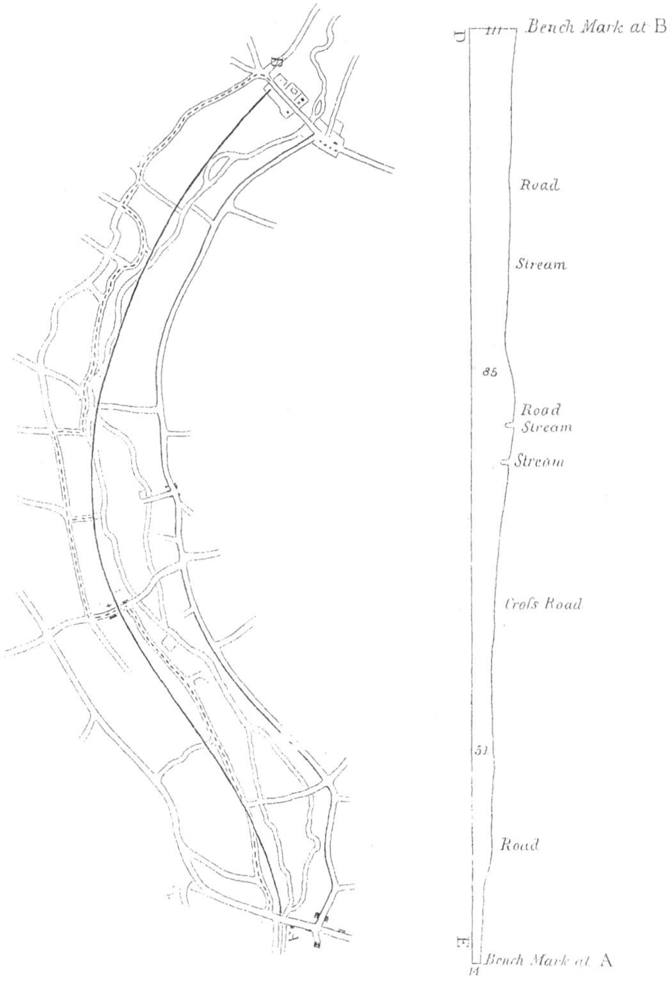

B

A

D

111 Bench Mark at B

Road

Stream

85

Road

Stream

Stream

Cross Road

51

Road

E

Bench Mark at A

14

SIMMS LEVELLING.

D.Van Nostrand, Publisher, N.Y.

A TREATISE

ON THE

PRINCIPLES AND PRACTICE

OF

LEVELLING,

SHOWING ITS APPLICATION TO PURPOSES OF

RAILWAY ENGINEERING AND THE CONSTRUCTION OF ROADS, &c.

BY

FREDERICK W. SIMMS, F. G. S., M. INST. C. E.,

CIVIL ENGINEER.

FIFTH EDITION, REVISED AND CORRECTED.

WITH THE ADDITION OF

MR. LAW'S

PRACTICAL EXAMPLES FOR SETTING OUT RAILWAY CURVES.

With three Lithograph Plates and several Wood-Cuts.

NEW YORK:

D. VAN NOSTRAND, PUBLISHER,

23 MURRAY STREET AND 27 WARREN STREET.

1870.

CONTENTS.

PART I.

ON THE PRINCIPLES OF LEVELLING.

PART II.

THE PRACTICE OF LEVELLING.

PART III.

COMPUTATION OF EARTH-WORK, ROAD-MAKING, THE CLINOMETER, ETC.

TABLES.

A TREATISE ON LEVELLING.

PART I.

ON THE PRINCIPLES OF LEVELLING.

LEVELLING is the art of tracing a line at the surface of the earth which shall cut the directions of gravity everywhere at right angles. If the earth were an extended plane, all lines representing the direction of gravity at every point on its surface would be parallel to each other; but, in consequence of its figure being that of a sphere or globe,* they everywhere converge to a point within the sphere which is equidistant from all parts of its surface; or, in other words, the direction of gravity invariably tends towards the centre of the earth, and may be considered as represented by a

* The figure of the earth is not exactly that of a sphere, but of an oblate spheroid flattened at the poles; the length of the equatorial diameter being 7924 miles, and that of the polar diameter 7898 miles. For our present purpose, it is sufficiently correct to consider it as a sphere.

plumb-line when hanging freely, and suspended beyond the sphere of attraction of the surrounding objects.

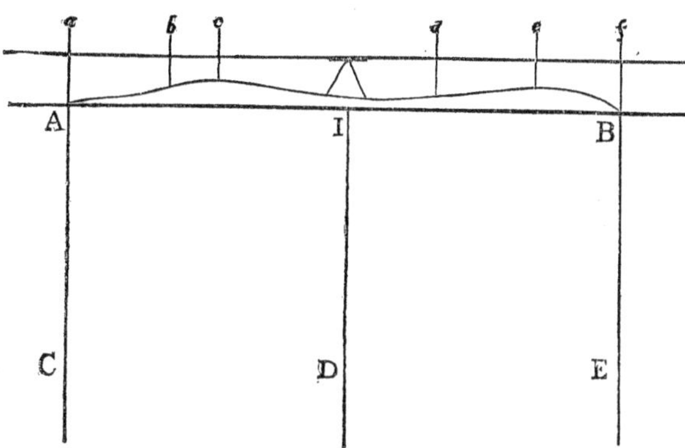

In the above diagram let the *straight* line A B represent the surface of the earth, upon the supposition of its being an extended plane, the direction of gravity at the points A, I, and B, would be represented by the lines A C, I D, and B E, all parallel to each other, and at right angles to the horizontal line A B. Now if the surface was undulatory, as shown by the curved line A B, and it was required to make a section representing it, an instrument capable of tracing out a line parallel to the horizontal line A B (as a spirit level), might be set up anywhere on the surface, as at I, and staves being placed or held along the line, as at *a*, *b*, *c*, *d*, &c., the different heights above the ground where such staves were intersected by the line so traced out, would at once show the relative level of all those points, with regard to the horizontal line, as a datum or standard of comparison.

But as the earth is a globe, its circumference must be circular, as I K L in the annexed figure ; the straight line A B will therefore not represent the surface of the

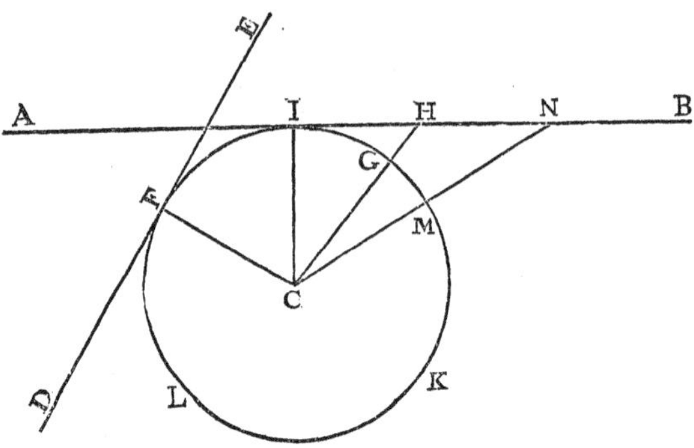

earth, but the sensible horizon of an observer stationed at the point I, to which point it is a tangent, being at right angles to the radius of the circle (or semi-diameter of the earth), I C. A line which is parallel to the sensible horizon of the observer, is the line traced out by our spirit-levels ; it is parallel to a tangent to the earth's surface at that point only where the instrument is set up,—thus A B is a tangent at I, and D E a tangent at F ; such being the fact, the difference of level between any two points cannot be determined by simple reference to a horizontal line, since every point on the surface of the globe (however near to each other) has a distinct horizon of its own.

If the earth were everywhere surrounded by a fluid at rest, or that its surface was smooth, regular, and uniform, every point thereon would be equally distant from the centre; but in consequence of the undulating form of the surface, places and objects are differently situated, some further from, and others nearer to, the centre of the earth, and consequently at different levels. The operation of levelling may therefore be defined as the art of finding how much higher or lower any one point is than another, or, more properly, the difference of their distances from the centre of the earth.

Referring to our last figure, we have seen that the line A B is a true horizontal or level line at the point I, but being produced in the direction A or B, rises above the earth's surface; and although it may appear to be level as seen from I, yet it is above the true level (which is represented by the circumference of the circle) at every other point, and continues to diverge from it the further it is produced; at G, the apparent line of level, as the horizontal line A B is called, is above the true level, by the distance G H, and at M by the distance M N, *the difference being equal to the excess of the secant of the arc of distance above the radius of the earth.*

The difference, G H, or M N (see last figure), between the true and apparent level may be thus found: put t in the adjoining diagram for the tangent I H, r for the radius C I of the earth, and x for G H, the excess of the secant of the arc of distance above the radius; I H being considered as equal to I G; then

$$(r+x)^2 = r^2 + t^2$$
$$\text{or } r^2 + 2rx + x^2 = r + t^2$$
$$\text{and } 2rx + x^2 = t^2$$
$$\text{or } (2r + x)x = t^2$$

But because the diameter of the earth $2r$ is so great with respect to the quantity (x) sought, at all distances

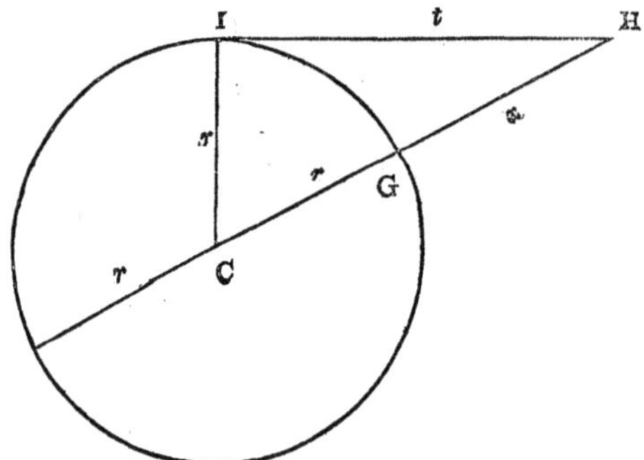

to which a common levelling operation usually extends, that $2r + x$ without sensible error may be replaced by $2r$, we then have

$$2rx = t^2$$
$$\text{and } x = \frac{t^2}{2r}.$$

Or in words: *The difference* (x) *between the true and apparent level is equal to the square of the distance* (t^2) *divided by the diameter of the earth* ($2r$), *and consequently is always proportional to the square of the distance.*

The mean diameter of the earth is 7916 miles, and the excess of the apparent above the true level for one mile $\frac{t^2}{2r} = \frac{1}{7916}$ of a mile, or 8.004 inches. At two miles, it is four times that quantity, or 32.016 inches; at three miles, it is nine times that quantity, or 72.036 inches; and so on increasing in proportion to the square of the distance. If we reject the decimal .004, and assume the difference between the true and apparent level for one mile to be exactly eight inches, or two-thirds of a foot, there arises the following convenient form for computing the correction of level due to the curvature of the earth, for distances given in miles, which may easily be remembered:

$$\text{correction} = \frac{2\,D^2}{3},$$

D being the distance in miles. Or in words: *Two-thirds of the square of the distance in miles will be the amount of the correction in feet.*

Example.

From a point on the Folkstone road, the top of the keep of Dover Castle was observed to coincide with the horizontal wire of a levelling telescope when adjusted for observation, and therefore was apparently on the same level; the distance (D) from the instrument to the Castle was four miles and a-half; consequently,

$$D^2 = 20.25$$

$$2\,D^2 = 40.50$$

$$\frac{2\,D^2}{3} = 13.5 \text{ feet, the correction required.}$$

From this it appears, that the keep of Dover Castle was 13.5 feet higher than the centre of the telescope on the Folkstone road ; but on account of the curvature of the earth, it was apparently depressed to the same level.

But the effect of the earth's curvature is modified by another cause, arising from optical deception ; namely Refraction. An object is never seen by us in its true position, but in the direction of the ray of light which conveys the impression or image of the object to our senses. Now the particles of light, in traversing the atmosphere, are, by the force of superior attraction, refracted or bent continually towards the perpendicular, as they penetrate the lower or denser strata ; and consequently they describe a curved track, of which the last portion, or its tangent, indicates the apparent elevated situation of a remote point. This trajectory, suffering almost a regular inflexure, may be considered as very nearly an arc of a circle, which has for its radius seven times the radius of our globe ; in consequence of which, the distance at which an object can be seen by the aid of refraction, is to the distance at which it could be seen without that aid, nearly as 14 to 13, the refraction augmenting the distance at which an object can be seen by about a thirteenth of itself. Hence, to correct the error occasioned by refraction, it will only be requisite to diminish the effects of the earth's curvature, or height of the apparent above the true level, by one-seventh of itself. Thus for our example of Dover Castle, $\frac{1}{7}$ of 13.5, or $\frac{13.5}{7} = 1.93$ feet nearly, to be substracted from 13.5, which leaves 11.57 feet for the height of Dover Castle above the level of a ce point on the Folkstone road.

The following Tables show the reduction of the parent to the true level, both for the curvature o earth only, and also for the combined effects of cu ture and refraction. The first gives the correc corresponding to distances expressed in miles, an second for distances in chains.

Table of the Difference of the Apparent and True Lev Distances in Miles.

Distance in Miles.	CORRECTION.			
	Curvature.		Curvature and Refraction.	
	feet.	inches.	feet.	inches.
¼	0	0.5	0	0.4
½	0	2.0	0	1.7
¾	0	4.5	0	3.9
1	0	8.0	0	6.9
2	2	8.0	2	3.4
3	6	0.0	5	1.7
4	10	8.1	9	1.8
5	16	8.1	14	3.5
6	24	0.1	20	7.0
7	32	8.2	28	0 2
8	42	8.3	36	7.1
9	54	0.3	46	3.7
10	66	8.4	57	2.1
11	80	8.5	69	2.1
12	96	0.6	82	3.9
13	112	8.6	96	7.4
14	130	8.8	112	0.7
15	150	0.9	128	7.6
16	170	9.0	147	2.3
17	192	9.2	165	2.7
18	216	1.3	185	2.8
19	240	9.4	206	4.7
20	266	9.6	228	8.2

Table of the Difference of the Apparent and True Level for Distances in Chains.

Distance in Chains.	CORRECTION.	
	Curvature in decimals of feet.	Curvature and Refraction in decimals of feet.
1	.000104	.000089
2	.000417	.000358
3	.000938	.000804
4	.001668	.001430
5	.002605	.002233
6	.003752	.003216
7	.005107	.004378
8	.006670	.005717
9	.008442	.007236
10	.010422	.008933
11	.012610	.010809
12	.015007	.012863
13	.017613	.015097
14	.020427	.017509
15	.023450	.020100
16	.026680	.022869
17	.030120	.025817
18	.033767	.028943
19	.037623	.032248
20	.041687	.035732
21	.045960	.039394
22	.050442	.043236
23	.055132	.047259
24	.060031	.051455
25	.065137	.055832
26	.070452	.060388
27	.075975	.065121
28	.081708	.070036
29	.087648	.075127
30	.093798	.080399

The correction for distances greater than those given in the latter Table may be computed by the following rule, the same by which the Table itself was computed :

Rule.—*To the arithmetical complement of the logarithm of the diameter of the earth, or* 2.3788603, *add double the logarithm of the distance in feet, the sum will be the logarithm of the correction for curvature in feet and decimals ; from which if one-seventh of itself be subtracted, the result will be the combined correction for curvature and refraction.*

The practice of levelling is one of the most delicate operations that fall within the province of a surveyor, requiring the utmost possible circumspection to avoid the numerous sources of error to which he is liable. More especially, as it is seldom possible for him, after levelling over a long tract of country, to conjecture in what portion of the work his error lies, if he should then find that he had been so unfortunate as to commit any, and, not unfrequently in such cases, sufficient time cannot be spared to go over the ground again ; as, for instance, when a section is required within a very limited time to produce before a Parliamentary committee, either to support or oppose any measure submitted to their consideration. We have witnessed an instance where such a committee, during their inquiry into the merits of a certain proposed line of railroad, had brought before them a *rival contemplated line* with *pretensions* to *great superiority ;* but it had been so hastily surveyed, that the learned counsel who had the supporting of the measure, acknowledged, in his opening address, that a trifling error at some unknown part of the line had been detected, which did not exceed fifty feet. We hardly need add, that the rival line was rejected.

The importance of extreme accuracy may also be felt, when it is known that from the section, the engineer has to make his calculations of the quantity of earthwork, in cuttings and embankments, necessary to carry into execution the intended measure, whether of a canal, a railway, or turnpike road, and of course the accuracy of the estimated expense is involved in it; and further, the fitness of the ground itself for such works is determined from the section; that is, whether the inclinations, which the undulations of the ground admit of being introduced, are suitable for the purpose either of a railway or turnpike road. And if the object be the formation of a canal, the section must show what extent of lockage will be required; not only affording a key to the expense, but also the possibility of its execution. We do not throw out these suggestions to alarm the mind of the young beginner, by bringing before him a fearful responsibility, but that he may understand the ultimate object of his labors, and to induce him, by carefulness and attention, to merit that confidence which is sure to be reposed in those who are known to possess such habits.

LEVELLING INSTRUMENTS.

It is essential to the good execution of work, that the surveyor should possess instruments most proper for the purpose, and of the best construction. Upon the subject of instruments, we shall generally refer the reader to a cheap work, entitled, "A Treatise on the principal Mathematical Instruments employed in Surveying, Levelling, and Astronomy, explaining their construction, adjustments, and use ;" where the various kinds of spirit-levels, and levelling staves, together with the method of performing their several adjustments, &c., are minutely detailed, and represented by engravings ;* and as the work alluded to contains also a similar account of the most important instruments used in surveying and astronomy, and has had an extensive sale, we presume it to be in the hands of most beginners in the profession ; we shall, however, give some particulars in this place, and annex a description of the cause of, and a remedy for, the *parallax* between the wires of a levelling telescope, and the levelling staves, which is the cause of much annoyance to observers.

SPIRIT LEVELS.

The Y *level*, so called from the supports in which the telescope rests, resembling in shape the letter Y, is the

* Also a work published by Mr. Weale, on Drawing Instruments, with Instructions for Field Work, in 12mo, price 3s. 6d.

oldest construction of the spirit-level now in use; its adjustments are convenient to be performed, but, on the other hand, this kind of instrument seldom retains its adjustments perfect for any length of time; besides, there are conditions in its construction which are assumed to be perfect, but which practical men know to present difficulties in the manufacture. The use of this instrument is now very much superseded by those of modern construction.

Troughton's Improved Level.—This instrument has been a very general favorite among engineers for a length of time; its construction renders its adjustments much more permanent than those of the Y level, and it is altogether a more stable instrument. The telescope, which, in the former instrument, is capable of reversion on its supports, for the adjustment of the line of collimation, is, in Troughton's construction, firmly fixed in its place, as is also the glass tube of the spirit bubble. The verification and correction of the adjustments are performed very differently, and may at first appear more complex and difficult than those of the other; yet when a person has once mastered and become familiar with his instrument, these apparent difficulties vanish.

The Dumpy Level.—This modification of the spirit-level has but recently been introduced by William Gravatt, Esq., and bids fair to become the favorite instrument among civil engineers. In its general figure it does not differ very essentially from the level last spoken of, but it possesses many decided advantages.

The aperture of the object glass is much larger for the same length of telescope; consequently more rays of light are admitted to the eye, producing the advantages of greater distinctness. We lately tried a *fourteen-inch* level, constructed upon Mr. Gravatt's principle, and found that we could distinctly read the levelling-staff at twenty chains (a quarter of a mile) distant, which was the utmost we could do with a *twenty-inch* level upon the old construction; we have, therefore, the advantage of a more portable instrument, fourteen inches in length, capable of performing the same work as a more cumbersome one of twenty inches. Besides this advantage, the instrument in question is more complete in its details. It possesses a cross level, placed at right angles to the principal level, which affords very great facility in setting up the instrument, and adjusting for observation, as will be hereafter described; it likewise has a reflecting mirror, mounted with a hinge joint, and capable of being placed on the principal level tube and adjusted, to show the observer if the instrument shifts from its horizontality whilst he is noting the observation; it also possesses other important though minor additions, all of which, in fact, could be applied by the maker to the other kind of instruments, if ordered, and for the particulars of which we refer to the work before alluded to.

From the large aperture and short focal length of the telescope, the instrument has altogether a dumpy appearance, and hence it is generally known by the cognomen of "Gravatt's Dumpy Level;" usually of nine or fourteen inches. We have seen some beautiful specimens of this kind of levelling instrument constructed for I. K. Brunel, Esq.

LEVELLING STAVES.

In the Treatise on Mathematical Instruments will be found a description of the different kinds of levelling staves in use. The former construction, even as improved by Troughton, was decidedly defective in practice, inasmuch as the staff had to be read off by the assistant, who had then to communicate the result to the observer; or, if he was not sufficiently intelligent to be intrusted with so responsible a duty, he was obliged, after the observation was made, to carry the staff to the observer, or wait for him to come and read off the height of the vane, and register it in his field-book. This occasioned great loss of time and uncertainty in the results, for the vane on the staff might possibly be shifted in the mean time. We remember an instance of an ignorant attendant holding the staff upside down, which at once introduced an error of several feet in the result. To obviate this, a new staff has been contrived, originally, we believe, by Mr. Gravatt, and subsequently by Mr. Hennett, Mr. Bramah, Mr. Sopwith, &c., each varying the mechanical arrangements, but all agreeing in retaining the main advantage, viz., a sufficiently distinct graduated face for the observer to read off the quantities himself through the telescope of his instrument; the sliding vane is therefore dispensed with, and the only dependence to be placed on the staff-holder is, that he may hold it perpendicularly. To assist him in this, a small plummet is suspended in a groove cut out in the side of the staff, by which its verticality can be determined in one direction, and the observer himself can detect if it be held aslant in the other direction, as may be understood from the diagram at page 26, which represents the staff *e* as it appears in the field of the telescope, which shows objects inverted. If the staff be held perpendicularly, it will appear between and equally distant from each of the two vertical wires *c d*, fixed in the telescope; consequently, if it be held aslant, it will cross the wires obliquely, and any want of verticality in the staff will be immediately detected, and the observer must signal to the staff-man accordingly. The advantages from the use of the modern staves, over those of the old construction, are so great, especially in saving of time, that we have no doubt of their general adoption.

THE IRON TRIPOD.

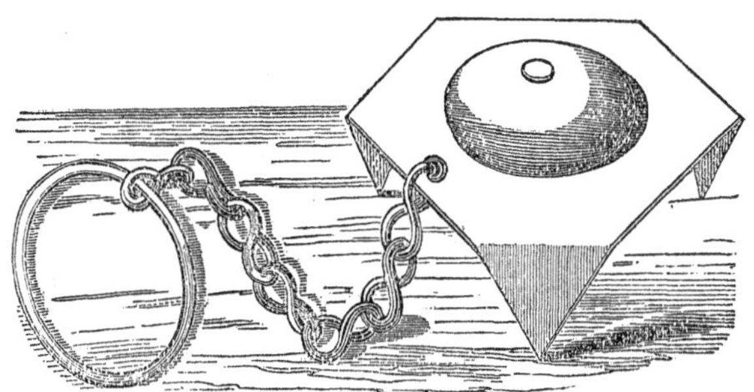

Another instrument of simple construction is represented in the preceding figure; its use is to rest the staff upon when held at any station. By this means the staff is sure to be kept on the same spot, and at the same height from the ground, while the observer is reading the staves both at the back and forward station on each side of the spirit-level; it is at present not generally used, but we consider it of more importance than is usually attached to it. It consists of a triangular piece of sheet iron, of about one-tenth of an inch in thickness, having the corners turned down to form the feet of the tripod, which are to be pressed into the ground by the foot of the staff-holder; a rounded piece of iron is riveted on the upper surface, to present a clean spot to rest the staff upon when held at the station; the chain with the attached ring is for the convenience of the staff-holder in lifting it from the ground, and carrying it from station to station.

THE MEASURING CHAIN.

In levelling operations it is in most cases necessary to note the relative distances of the staves from each other, from the spirit-level, or from some given point or place, otherwise no section of the ground levelled over can be made. For this purpose a measuring tape may be employed where the distances are short, but in most cases the means employed is a chain; the one commonly used is 4 poles in length, called Gunter's chain, which is divided into 100 links of 7.92 inches each. In many cases, however, this will not be found so convenient as the use of a chain with links of 1 foot in length; but there is a practical inconvenience attending these long links where the ground is rough and uneven, as the links are likely to get bent in being drawn through the hedges and rough places; whenever this occurs the chain is reduced in length, and, unless discovered and rectified, a considerable error in distance will very soon result. When we have had occasion to use such a chain over rough ground we have had the links made 6 inches long, and although it occasioned more trouble in noting and registering the distances, yet the liability of the links to become bent was greatly diminished. No measurements are required in taking what are called running or check levels, the object of which is merely to test the accuracy of a section previously made, by finding the difference of level between certain points on the section, to see if the results are identical with the former determination; which is the same thing as acertaining the whole difference of level between distant places. Neither are any measurements required to produce a section if you possess a correct map or plan of the district or line, for if the level points are noted on the said plan, their relative distances can be taken therefrom by its scale; this, however, can only be considered as an approximate operation as far as the horizontal measures are concerned. In this way, however, many extensive trial sections for long lines of railway have been made by means of the Ordnance maps, and will, if properly done, determine the general features of the country sufficiently for the engineer to choose the best route for a minute and detailed survey, which would cost too much time and money to undertake in the first instance where there exists any doubt between two or more routes as to which it would be most judicious to adopt.

ON INSTRUMENTAL PARALLAX.

The foregoing is an account of the instruments necessary for the purposes of levelling ; but before closing this part of our subject, we think it may be useful to add some particulars respecting instrumental parallax, which we have occasionally found to be the source of much annoyance to the surveyor. This has invariably arisen from ignorance of the principles of the telescope, and hence, not knowing how the parallax arises, the means of removing it have not been understood ; we shall endeavor to explain, in a popular manner, both the cause and the remedy.

The rays of light which proceed from surrounding objects, and which, by entering our eyes, convey to us the sense of vision, move in perfectly straight lines, unless turned from their rectilineal course by the intervention of a refracting or reflecting medium, and whatever portion of such rays as can enter our eyes may (without sensible error) be considered as moving not only in *straight*, but *parallel* lines ; the more remote the object is, the more nearly this will be the case. In the adjoining diagram, let A B represent the section of a

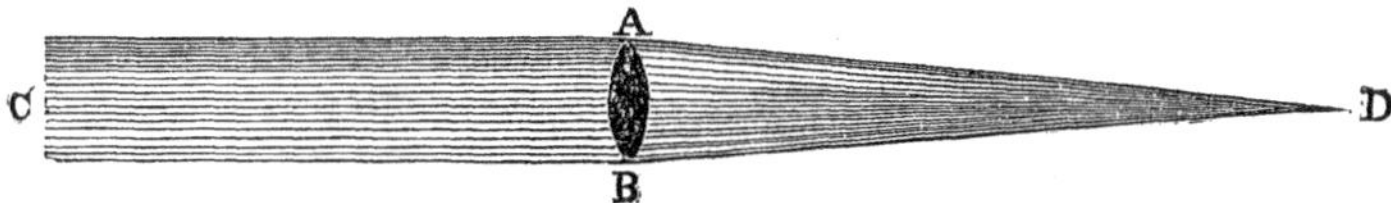

lens (or object glass of a telescope) ; let the parallel lines on the left represent the rays of light coming from some distant object in that direction ; the instant they impinge upon the glass, and in passing through it, they suffer refraction—that is, they are bent out of their former rectilineal path—and on leaving the lens at the opposite side, they converge to a certain point D, which is the focus of the object glass (in this point *all* the rays passing through a *perfectly formed* glass meet, and it is situated on the line C D, the direction of the ray which passes through the centre of the glass, the only one that continues its former course, and is called the axis of the lens) ; the concentration of the rays form an image of the distant object in the focal point D, "and if a piece of ground glass, transparent paper, or a plate of glass having one surface covered with a dried film of skimmed milk, be held up at D, a person looking at it from a few inches behind would see a perfect image of the distant object formed on the ground glass ; and by steadily keeping the eye in the same position, the ground glass may be removed, and the image will appear in the same spot suspended in the air."

Now let us imagine the lens applied to the construction of a telescope, and the adjoining diagram to represent a section of it; the image of a levelling staff held at a distance, in the direction of C, would be formed at the point W, the focus of the object glass; let D F represent the eye-glass, which is fixed in a sliding tube, and together called the *eye-piece*. The eye-piece may be considered as a microscope, with which

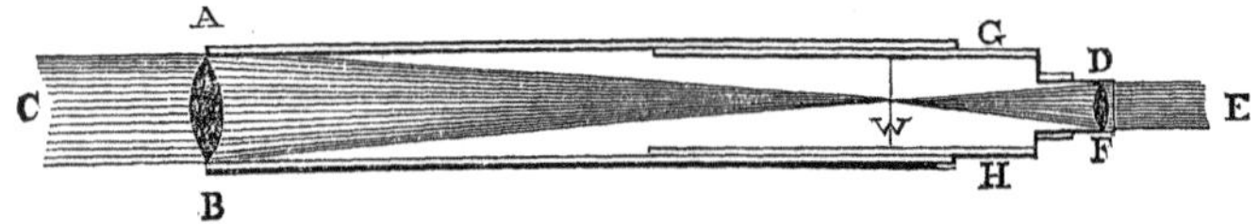

the observer magnifies the image of the object formed at W; to do this, it will readily appear to the reader that its distance from the image at W must be such as to cause its focal point to coincide therewith, making that point the common focus of the two glasses; for the purpose of effecting this, the eye-piece is made to slide either in or outwards, and the observer can tell when it is at the proper distance, for he will then obtain a perfectly distinct view of the object. The axis of the two glasses forms a continued straight line C E, which in a telescope is technically termed the optical axis of the instrument, or line of collimation; this imaginary line is, in levelling telescopes, the zero, from whence the readings on the staff are taken. It is therefore necessary to represent it by something tangible, that shall at the same time not interfere with the rays of light passing through the telescope to the eye; this is done by fixing across the interior of the telescope very fine wires, or threads from a spider's web, so that their intersection may not only coincide with the axis C E, but cross it precisely at W, the common focus of the two glasses, where the image of the staff (or distant object) is formed, and therefore the wires and the staff will appear to an observer as one object, or, at least, equally distant from him. The following diagram shows the appearance of the wires and the staff as seen through an inverting telescope; where *a b* represents the horizontal wire, *c* and *d* two

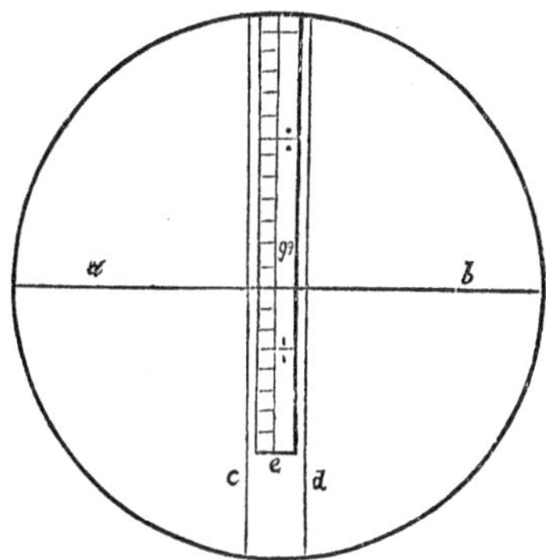

wires placed at right angles to it, and separated so as to admit, at usual distances, the staff *e* to appear between them, by which the observer can always tell if the staff-man holds it erect in a lateral direction, as before explained. The staff is represented as seen at the moment of completing an observation; the horizontal cross wire coinciding with the division .20 above 16 feet, the staff being read downwards in consequence of its apparent inversion; the reading, therefore, of such an observation, to be entered in the field-book, would be 16.20 feet.

The adjustment of the line of collimation consists in making the centre of the horizontal wire (or intersection of the wires in instruments intended for measuring angles) coincide with the optical axis of the telescope ; this, when once accomplished, will, with care, keep correct for a long time, but the placing it in the common focus of the two glasses requires attention at every observation. For detailed instructions upon the former, we refer to the treatise on Mathematical Instruments, &c. ; but as the latter forms part of every observation, and is the source of the perplexing parallax, we shall speak of it in this place.

The cross wires are fixed to a plate, called a diaphragm, attachedby screws to the slide G H, which also carries the slide D F of the eye-piece. The point W, or focus of the object glass, does not remain constant for terrestrial objects, but varies with every change in the distance of the staff; if it be brought closer to the instrument, the image, or focal point, will recede further from the glass, and *vice versa ;* therefore, the wires and the focus of the eye-piece must be brought to coincide with that of the object glass by their respective slides; and first, the eye-piece should be moved in its slide till its focus coincides with the wires in the tube G H; when this is accomplished, the observer will see the wires perfectly sharp and well-defined; next, motion must be given to the slide G H, by turning a milled head attached to the telescope, which gives motion to the slide by rack work ; this will carry both the wires and the focus of the already adjusted eye-piece to coinc de with the focus of the object glass, on whatever part of the optical axis of the instrument it may be situated. When this is done, the adjustment of the telescope for observation will be complete, and its proof consists in the observer having at the same time a clear and well-defined image both of the staff and the cross wires, which will be the case if they seem to be *attached* to each other, or, in other words, appear equally distant from him ; and the moving about of the observer's eye does not detect any apparent displacement of the staff, with respect to the wires. Such a displacement, or relative motion, is what is meant by *parallax ;* and when it exists, it must be got rid of by a repetition of the adjustment of the glasses as above described, till the motion of the eye will no longer detect the least apparent movement, or passing and repassing of the wires and the staff; till this is done, no correct observation can be made.

From what has been advanced on the subject of the corrections for curvature and refraction, it may be necessary, before entering upon any practical examples, to remark, that such corrections are very seldom applied in practice, the observer, by the arrangements of his operations, doing away in a great degree their injurious effects, which we will endeavor to explain.

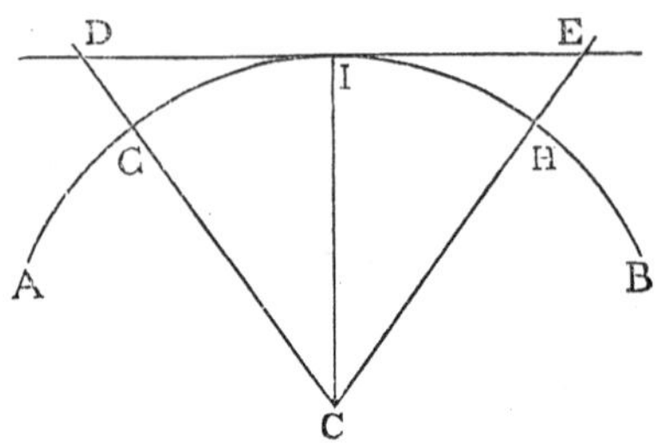

Suppose it were required to find the difference of level between any two points G and H in the preceding figure ; let A B represent a portion of the earth's surface, let C represent the centre, and C G, C I and C H the radii of the earth. Now a spirit-level being set up and adjusted at I, an observer looking through the telescope would see objects in the direction of the horizontal line D E only, and a staff held upright at H would be read off in the point E on the horizontal line ; but this point is higher than the true level by the distance H E, which is the correction for curvature due to the distance I H (see page 9) ; and if that quantity be subtracted from the reading of the staff, the remainder will show the difference of level between the points I and H. If the same process be gone through by holding a staff at G, then the difference of level between G and I will also be ascertained, which being compared with the former difference, will show how much higher one of the points G or H is above the other ; but it must be evident, that if G and H be equally distant from I, the horizontal line D E, being a tangent to the surface at the middle point I, must cut the staff at D on the same level with the point E; that is, C D is equal to C E, therefore D and E are level points, being equidistant from the centre of the earth; and if the reading of one staff above the ground is greater than the reading of the other, the difference will at once show the variation of level between the points where the staves were held, viz., G and H; the effect of curvature is thus removed by *simply placing the instrument midway between the station staves.* The effects of the atmospheric refraction will likewise be done away with in the same process, because it will affect both observations alike, unless under peculiar circumstances of the weather, &c., over which the observer has no control.

The above method of finding differences of level, by placing the instrument as near as possible midway between the two staves, and noting their readings, is the one adopted in practice; but as it can scarcely ever happen, on account of the extent of the work, that one placing of the instrument will complete it, a succession of similar operations must be performed, as shown in the annexed engraving.

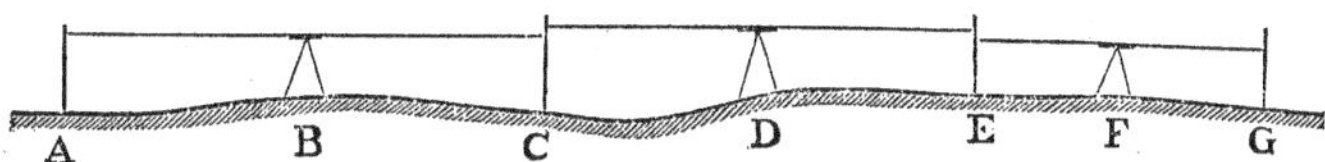

Suppose it were required to find the difference of level between the points A and G; a staff is erected at

A, the instrument is set up at B, another staff at C, at the same distance from B that B is from A. The readings of the two staves are then noted; the horizontal lines connecting the staves with the instrument represent the visual ray or line of sight. The instrument is then conveyed to D, and the staff which stood at A is now removed to E, the staff C retaining its former position, and from being the forward staff at the last observation, it is now the back staff; the readings of the two staves are again noted, and the instrument removed to F, and the staff C to the point G; the staff at E retaining the same position, now becomes in its turn the back staff, and so on to the end of the work, which may thus be extended many miles; the difference of any two of the readings will show the difference of level between the places of the back and forward staff; and the difference between the sum of the back sights and the sum of the forward sights will give the difference of level between the extreme points; thus:

	Back sights. ft. dec.	Fore sights. ft. dec.
A and C................	10.46	11.20
C " E................	11.33	8.00
E " G................	7.42	7.91
Sums.....................	29.21	27.11
	27.11	
Difference of level.........	2.10	

showing that the point G is 2 feet and $\frac{1}{10}$ higher than the point A.

The foregoing process is called compound levelling. The following is an example of simple levelling, being performed at one operation, and therefore subject to the correction for curvature and refraction to obtain a correct result.

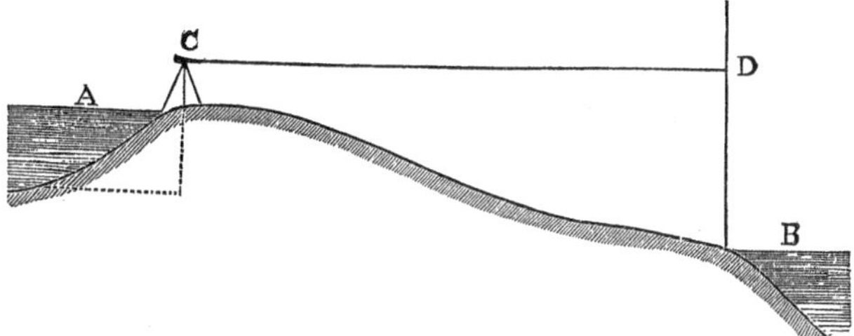

Suppose it were required to drain a pond and marsh A, by making a cut to a stream at B, a distance of thirty chains; let a level be set up at C, and directed to a staff held upright at the edge of the water at B. The horizontal line C D represents the line of sight which would cut the staff at D, the reading being 17.44 feet; the height of the instrument above the ground was 4 feet, and the depth of the pond 10 feet; therefore the difference of level between the bottom of the pond and the surface of the stream was as follows:

		ft. dec.
Reading of the staff......................		17.44
Height of instrument..............	4.00	
Depth of pond	10.00	
Curvature and refraction for 30 chains (see Tables, pages 12 and 13).....	0.09	
	——	14.09
Difference of level......................		3.35

PART II.

THE PRACTICE OF LEVELLING.

ON RUNNING OR CHECK LEVELS.

To present, in the clearest possible manner, the practical application of the principles of levelling, we propose describing some operations in detail. We shall, therefore, commence with a case of a simple kind, which will prepare the way for more complicated examples. When a section of a line of country has been completed (for any purposes whatever), it is in most cases necessary to check its accuracy by repetition; but in doing this, it is seldom requisite to level over precisely the same line of ground, unless there is cause to suspect its general correctness, but to follow the most convenient and nearest route, and at intervals to level to some known points on the exact line of section, which will give *their* differences of level; the points thus selected are generally what are called bench marks, and are nothing more than marks or notches cut upon gateposts, stumps of trees, mile or boundary stones, or any similarly immovable objects, contiguous to the line of section, and at frequent intervals. These bench marks are made by the person who takes the section in the first instance, and are sometimes previously determined upon. When the section is complete, their relative heights with regard to the base line or datum of the section become known ; consequently, they may be considered as so many zero or fixed points on the line, easily recognizable, from whence any portion of the work may be levelled over again ; or branch lines of level may be conducted in any direction, and the levels of such branches be comparable with those of the main line.

When, in checking the principal levels, by proceeding in the most convenient direction from bench mark to bench mark, it is found that the differences of level prove identical with those on the section, or within the limits of probable error, it may be presumed that all the intermediate heights are likewise correct ; it is, however, just possible that equal errors of an opposite kind may have been committed, when, the *sum* of each being of the same magnitude, a balance of errors would cause the extreme points to be right, whilst the intermediate levels would be incorrect ; but the probability is so much against such an occurrence, that we believe, unless there be some particular reasons for so doing, the whole exact line of a section is seldom levelled a second time for the purpose of checking the former results only.

From what has been remarked, it will appear evident that in taking running or check levels, there is no necessity for the use of the chain, or the compass attached to the instrument, the distances and bearings having all been determined at the time the principal levels were taken.

The example we are about to give of this kind of operation is represented in the engraving, Plate I., which shows both the ground plan and the section. The strong black line on the plan is that of the section to be checked, and proceeds from a bridge in the town of A, in a circuitous direction along a valley, and nearly parallel to the course of a river, to a bench mark in the town of B ; this originally formed a portion of a more extensive survey. We have selected this portion of the line as explanatory of our present subject ; the route taken in proving the work is represented by the dotted line, and was confined to the public roads, that being considered the most convenient, because it would altogether exclude the necessity of passing through private property, as the surveyor would most likely have been ordered off, a great feeling of opposition existing among the owners and occupiers of the said lands ; and further, the public road crossed the line several times, by which a number of intermediate points could be checked. Before giving the particulars of this example, we shall explain in detail the method of conducting the necessary observations.

In the first instance the staff-holder must place his staff on the bench mark from whence the levels are to commence. (In the case of our example the staff was first placed on a peculiarly shaped stone on the crown of the bridge at A, which could easily be recognized from description at any future time, if ever it should be necessary to refer to this spot again; it therefore answered as a bench mark.) The surveyor must next set up his spirit-level in the most suitable spot which presents itself, from whence he can have an uninterrupted view, not only of the staff at the back station, but also for a considerable distance in the direction he wishes to carry his levels. The station selected should not in any case exceed four or five chains, and if it be only half that quantity, there will be less likelihood of error; for when long sights (as they are usually termed) are taken, unless both the back and forward stations are equally distant from the instrument, errors will gradually creep in upon the results, which, in a long series of levels, are liable, by their accumulation, to become of serious consequence. The proper station being determined upon,* and the tripod legs of the instrument spread out and thrust into the ground sufficiently to insure its stability, the observer must adjust his level for observation in the following order:— First, he must draw out the eye-piece of the telescope till he sees the cross wires perfectly well defined; then, directing it to the staff, he must turn the milled-headed screw, on the side of the telescope, till he can likewise distinguish with the utmost possible clearness the

* It must be borne in mind, when we thus minutely detail what may appear to the practical man as naturally obvious, that we are writing for the information of those who have never had any practice whatever.

smallest graduations on the staff; that these two adjustments be very carefully and completely performed, is of more consequence than is generally supposed, for upon them depends the existence or non-existence of parallax. If any parallax is detected, it must be removed, or the observations will be incorrect; its existence may be detected by the observer moving his *eye* about at the same time that he is looking through the telescope at the staff; and if he sees that the cross wires do not appear to have the least motion with regard to the divisions with which they are coincident, then no parallax will exist; but if any motion appears to take place between the wires and the staff, it is a proof that one or both of the foregoing adjustments have been imperfectly made.

To remedy this inconvenience the eye-piece should first be moved to try and improve the distinct appearance of the cross wires. The observer will be greatly assisted in this operation if he holds a sheet of white paper before the object glass, which, at the same time that it prevents other objects from attracting his attention, presents a clean white disk, or ground, for the wires to be seen upon; and when he is satisfied that they are as sharp and well-defined as possible, he must repeat the movement of the milled head by the side of the telescope till he is equally satisfied of the distinct appearance of the graduations on the staff; then let him again move his eye about before the eye-glass to see if any parallax still exists, and if so, he ought to repeat the above simple operation until it is removed. We have known the parallax of a telescope to be a source of great annoyance to persons in the profession, which has led us to be thus minute upon what to some would appear very simple. We have for the like reason given an explanation of its nature, &c., at page 23.

The turning the mill head to obtain distinct vision of the staff, in the old construction of instruments, communicated motion to the object glass; but in those of recent contrivance, it moves the whole of the eye end of the telescope, and with it the cross wires. In either case, the distance between the object glass and the wires is increased to a proper extent; the modern contrivance appears to be the most approved. The adjustment of the eye-piece for distinct vision when once made, is not likely to require alteration the whole day, unless it be accidentally deranged; but that of obtaining distinct vision of the distant staff (together with the one we shall next describe) must be performed at every station, as it varies with the distance of the staff, as explained at page 27.

Having made the above adjustments perfect, bring the spirit-bubble into the centre of its glass tube, which position it must retain unmoved in every direction of the instrument; or, in other words, the bubble must indicate a true level during the time the telescope is turned completely round horizontally on its staff head; this is accomplished by bringing the telescope successively over each pair of the parallel plate screws, and giving them motion, screwing up one while unscrewing the other to a corresponding extent ; but if the telescope is supplied with a cross level, as in that contrived by Mr. Gravatt, the two bubbles, being at right angles to each other, will at once show which pair of screws require turning, in order to produce an indication of level in both bubbles. In the Treatise on Mathematical Instruments there is given an ample explanation of the adjustment of levels in all their details; upon such subjects we shall once for all refer to that work.

Having adjusted the level for observation, it must be directed to the back staff, of which a clear view must be had ; then note with all possible exactness the foot, and decimal fraction of a foot, with which the central part of the horizontal wire appears to be coincident, which enter in the proper column of the field or observation book. This column should be headed "Back Sight," or "Back Station," as in the example given at page 45. As soon as it is registered, look to see that the spirit-bubble has not removed from its central position, and then repeat the observation, to insure that no mistake had been made in noting it ; this should be invariably done, to guard against errors.

The back observation being made, turn the telescope round in the forward direction, and obtain a distinct view of the staff, by turning the milled head at the side of the telescope ; then look at the spirit-bubble, and if it has at all changed its position, by receding towards either end of its tube, bring it back to the centre by the parallel plate screws, as before described (this can be done so readily, and without moving the telescope, when a cross level is attached, and having likewise other advantages, that we recommend its universal application to spirit-levels) ; then, by looking through the telescope, observe what division on the staff is intersected by the cross wire, and enter the reading in the proper column of the field-book, which should be headed "Fore Sight," or "Fore Station." Having entered it, look to see that the bubble is still correct, and then verify the observation by noting it again, which will complete the first levels.*

It may be worth remarking that, in setting the level up, the pointed legs should be pushed into the ground sufficiently to insure the stability of the instrument, and likewise that the observer should move himself about the instrument, whilst taking the levels, as little as possible, taking care not to strike the legs with his feet. Caution in these matters is required, for sometimes the least movement of the person will derange the levels of the instrument, particularly on loose or elastic ground ; to do away the inconvenience arising from this source, a reflector has been contrived to fix on the top of the telescope tube, by which the observer can see both the staff and the reflected image of the

* When taking levels for the formation of a section, it is sometimes necessary to note the bearing of the compass needle, and to measure distances, as will be explained hereafter.

spirit-bubble at the same time, and then he can make his observation at the instant he sees the bubble in its proper position. The foregoing description of the method of taking levels is general, and applies equally to every kind of levelling operation, with whatever additional matters may require attending to, when taking levels for the formation of a section, &c., which we shall hereafter describe.

The first levels being completed, the surveyor must take up his instrument, and, passing the man who holds the forward staff, proceed to some convenient spot to set up the instrument a second time, which, as before remarked, should not be more than four or five chains distant; the other man, also, who held the staff at the back station, must likewise take up a new station still further onwards in the required direction, and as nearly as possible at the same distance from the instrument as the instrument is from the staff, which has now become the back station; it being in every case necessary, to insure correct work, that the instrument should occupy very nearly the middle point between the staves, for reasons which will be understood by those who have perused the former part of this book. Having set the instrument up, adjust it for observation as before, viz., see that the cross wires are distinct; turn the milled head by the side of the telescope till the graduation on the staff is quite distinct, and no parallax exists; and, lastly, set the spirit-bubble level in every direction of the telescope by the parallel plate screws; which done, note the reading on the back staff, and enter it in the book; then examine the bubble, and again read the staff to insure accuracy; then turn the telescope about, and do the same for the forward station, which will complete the second level. As the third and fourth, and all the following levels are conducted in precisely the same manner, it will be unnecessary to repeat the instructions again.

The man holding the back staff should be instructed never to move it in the least from its position till the forward observation is completed, which he can always tell by seeing the surveyor carry his level onwards. It is sometimes the practice to use one staff only, and after taking the back observation, to cause the assistant to go on and take up a position suitable for a forward station; but besides the loss of time attendant upon such a process, if the instrument should in the interval get moved by accident, those two observations will be incorrect, unless the back sight be taken again, and this cannot be done unless the precise spot before occupied by the staff can be identified, which is sometimes uncertain. When this is the case, no alternative is left but to go back and renew the work at the last bench mark, or known station; and if none such exist, the whole operation will probably have to be gone over again, where great accuracy is required.

The iron tripod, described at page 20, should in all cases be placed on the ground by the staff-holder, to rest the staff upon, as it insures to the observer the certainty of the staff keeping exactly the same spot when the face of it is presented to him in the two directions, forward and backward. The staff-holder should likewise be instructed to hold the staff perfectly upright, which he can himself determine, in one direction, by a little plumb-weight suspended in a groove in the staff; and as the observer can tell if he holds it upright in a lateral direction (as explained at page 20), he should frequently look to see if he signals for him to move the upper end of the staff to the right or left, taking care not to disturb its position on the iron tripod.

We have been supposing the use of the newly introduced staves, as we do not expect that those of the former construction will hold their ground against them, they having the advantage of providing to the observer the means of noting the reading of the staff himself. If, however, from habit or otherwise, the use of the staff with the sliding vane should be preferred, the foregoing instructions equally apply; the only difference in its use is, that the observer must signal to the staff-holder to move the vane up or down on the staff, till it appears bisected by the cross wires of his telescope; then the reading of the staff must be noted, and entered by the assistant in a temporary book carried by him for the purpose; or if he cannot be trusted to perform so important a part of the business, he must convey the staff to the observer, or wait for him to come and read it himself. It requires no comment to show the uncertainty, and loss of time, in this method of proceeding compared with the use of the newly-contrived staves.

Having explained the method of taking observations for checking levels, we must refer to our example. The levels, as before stated, were taken along the public road shown by the dotted line, that being the most convenient route from the town of A to the town of B, avoiding the necessity of passing through private property; the strong black line on the plan shows where the original section was taken; the section itself is represented above the plan, and is drawn to two scales; the one giving the horizontal measure, is the same as that of the plan, viz., one inch to one mile; and the vertical scale, $\frac{1}{4}$ inch to 100 feet; from this section it appears that the crown of the bridge at A is fourteen feet above the datum line D E of the section, and that the bench mark (a stone by the road side) at B is 111 feet above the same datum; therefore the difference of level between the two places is $111 - 14 = 97$ feet. Now, by referring to our observation book, of which we have subjoined a copy, we make the difference of level to be 96.8 feet, differing from the original section no more than two-tenths of a foot, or 2.4 inches, a quantity that may be disregarded; the inference, therefore, to be drawn from such a coincidence in the two results is, that the whole of the section between the points in question is sufficiently correct.

Copy of Field-Book, for running or check levels.

Back Sights.	Fore Sights.	Remarks.
0.34	3.16	Back Θ on B. M. on the bridge at A.
5.86	5.61	
4.19	4.24	Forward Θ at corner of road leading to B.
5.44	1.20	
4.96	3.20	
4.73	1.32	At crossing of line.
6.10	2.00	
5.33	3.96	
5.91	1.83	
5.70	0.90	
6.02	1.21	Staff placed on post notched for B. M.
1.21	4.00	At crossing of line.
3.53	6.07	
3.96	5.34	
3.94	4.81	
3.98	6.08	
4.08	4.94	Upon line.
3.90	3.96	
4.84	2.42	
1.54	5.12	
4.69	4.97	
5.04	1.60	
2.24	3.86	Upon line.
7.25	1.89	
4.03	1.30	
9.54	0.19	
6.70	1.70	
9.40	4.06	
6.44	0.38	
11.00	0.46	
5.98	1.30	
11.12	1.78	
9.84	2.20	
0.18	0.32	Upon line.
4.72	0.10	
8.89	0.77	
10.02	0.92	
10.00	1.03	
8.58	1.19	
9.53	1.18	
230.75 102.57	102.57	Sums.
128.18		Difference.

Copy of Field-Book—continued.

Back Sights.	Fore Sights.	Remarks.
128.18		Brought forward.
9.90	0.68	
9.04	0.35	
10.00	8.52	
3.00	11.55	
3.68	0.88	
7.21	8.75	
1.99	10.48	
0.65	10.00	
4.48	10.44	
1.47	10.30	
1.55	11.70	
2.45	9.88	
3.78	1.04	
6.64	2.65	Forward Θ on B. M. called B.
194.02	97.22	Sums.
97.22		
96.80	....	Difference=dif. of level betw'n A and B.

The back sights being greater in amount than the forward sights, it is evident that the bench mark at B was higher than the bench mark at A by the difference of the two sums.

LEVELS FOR THE FORMATION OF A SECTION.

Next to the running levels, the most simple case that can occur is, to take the levels of a line of country where the ground plan is already made, and the exact line of section determined upon, and in some instances picketed out. It is then only necessary, in addition to what is required for running levels, that the distance between the levelling staves, or the whole distance at every station from the starting point, be measured. The instrument should be placed, as usual, as near as can be at an equal distance from each staff; but it is not essential that it be placed in the exact line between them, unless it should happen to prove the most advantageous position. Plate II. represents an example of this kind of work, the survey of the land having been completed, and the plans of the fields, etc. drawn; the strong black line A B was the direction determined upon as the most suitable for a portion of an intended line of railroad, and the section was accordingly taken; a bench mark had been previously agreed upon at each extremity (A and B), from whence other surveyors could take up the levels, and carry them onwards in both directions.

First, a staff was placed on the bench mark at A for a back station, and another staff was held up for a forward station, in the adjoining field, but exactly on the line as marked down on the plan, a copy of which the surveyor had in his possession; the instrument was then set up, as near as could be estimated, or the level of the ground would admit, at an equal distance from each staff, so as to be able to read them both; the adjustment of the instrument for observation, as described at page 35, was carefully attended to; and the reading of the staves noted. As soon as the observations were made, the distance from staff to staff was measured with a Gunter's chain, which completed the first level.

The measurement of the distances can be more conveniently performed, and with a great saving of time, by two additional assistants, who can be measuring, whilst the surveyor proceeds to direct the man who held the back staff in the last case, to take up a forward station precisely on the line as laid down on the plan. The staff which was the forward station in the last case now becomes the back station, and the instrument must be set up so as to read both stations as before, and as nearly equidistant from them as can be ; by the time the instrument is adjusted, and both the staves read off, the assistants would have completed the measurement from the bench mark A to the first forward staff, and be ready to continue on to the second one ; whilst this is doing, the instrument and back staff can be carried forward and set up, &c., as before ; by a continued repetition of a similar process, the whole line A B was levelled.

The measuring assistant should report to the surveyor the total distance of each forward staff from the bench mark at A as soon as it is determined, or, if thought more convenient, he may keep a book to enter the distances in, which should be ruled in two columns, one for his distances, and the other for references to them; as *a*, *b*, *c*, etc., or the numbers 1, 2, 3, etc., placed opposite ; and if the observer makes similar notes in his book to each pair of sights, there can arise no mistake in placing the correct distances opposite the corresponding levels, when the measurer makes his return.

The following is a copy of the field-book of the

example given in Plate II.; showing the manner of keeping it, and also the method adopted of reducing the levels to obtain the actual heights of each station, with regard to the starting point, for the purpose of drawing the section ; which we shall then explain.

LEVELLING FIELD-BOOK.

Distances.	Rise.	Back Sight.	Fore Sight.	Fall.	Reduced Level.	Remarks.
519	5.83	13.71	7.88	......	+ 5.83	
1315	......	9.40	16.30	6.90	− 1.07	
1542	......	3.87	11.71	7.84	− 8.91	
1850	.. ...	2.63	12.41	9.78	− 18.69	
2358	13.67	14.62	0.95	......	− 5.02	
2698	15.55	17.00	1.45	......	+ 10.53	
3357	......	10.66	15.40	4.74	+ 5.79	
3758	......	2.87	17.00	14.13	− 8.34	
3976	.. ...	3.40	10.32	6.92	− 15.26	
5077	3.49	5.73	2.24	......	− 11.77	
5904	15.69	16.54	0.85	......	+ 3.92	
6124	15.19	16.08	0.89	......	+ 19.11	
6437	13.83	14.56	0.73	......	+ 32.94	
7467	......	10.36	14.06	3.70	+ 29.24	
8369	8.48	9.84	1.36	......	+ 37.72	
9303	2.80	9.80	7.00	......	+ 40.52	Centre of road at 215 links.
	......	2.30	10.96	8.66	+ 31.86	
9679	......	10.96	14.46	3.50	− 28.36	
9936	......	2.08	15.05	12.97	+ 15.39	
10164	......	1.75	16.58	14.83	+ 0.56	
10576	......	1.84	17.10	15.26	− 14.70	
11423	......	0.00	7.43	7.43	− 22.13	Forward θ at corner of Wood.
13066	1.88	5.38	3.50	......	— 26.25	
14954	4.00	8.50	4.50	......	− 16.25	
15650	3.94	5.30	1.36	......	− 12.31	
17345	0.80	10.20	9.40	......	− 11.51	
19135	6.46	6.86	0.40	.....	− 5.05	
19359	7.04	11.00	3.96	......	+ 1.99	
19631	8.27	11.80	3.53	.....	+ 10.26	
19841	7.85	10.53	2.68	......	+ 18.11	Forward θ at edge of Wood.
20561	6.84	8.22	1.38	......	+ 24.95	
21671	6.56	8.76	2.20	......	+ 31.51	
	......	14.00	14.50	0.50	+ 31.01	Road at 450 links.
22710	10.18	14.50	4.32	......	+ 41.19	
23221	8.14	9.14	1.00	......	+ 49.33	B above A.
Sums.	166.49	304.19	254.86	117.16		
	117.16	254.86				
	49.33	49.33				

The first column contains the measured distances from the starting point to every forward station expressed in links of Gunter's chain. The two central columns, headed "Back Sight" and "Fore Sight," contain the readings of the two staves at the back and fore observations respectively. The *difference* of such readings is placed in one of the two side columns headed "Rise" or "Fall," according as the ground at the forward station is higher or lower than that at the back station. If it be highest (or the ground rises, as it is called), then the forward reading will be the smaller of the two ; but if it be the lowest (or the ground falls), then the forward reading will be the greater of the two ; thus, in our first reading, the back observation is 13.71, and the forward observation 7.88, their difference= 5.83 feet, which is the difference of level between the two points ; and as the forward rea^ing was the smaller of the two, it is clear that the ground was rising at that place, and therefore the difference of the readings, viz. 5.83, is placed in the column of rises. In the next three successive pair of sights, the forward readings are the greatest, indicating a continued descent of the surface line, and the differences of those readings are inserted in the column of falls, viz. 6.90, 7.84, and 9.78. At the next following sight, the forward reading is again the smallest, therefore the difference 13.67 is placed in the column headed "Rise," and so on of the rest. No mistake can arise by placing the subtraction in the wrong column, as in every instance it must be placed in the column adjoining the larger quantity; thus if the fore sight is greater than the back sight, the resulting quantity must be placed in the column of falls, which is adjoining to that containing the reading of the fore sight, and *vice versâ.*

The adjoining column, headed "Reduced Levels," contains the absolute heights of each forward station above the datum line of the section, or a horizontal line passing through the starting point or bench mark A; these quantities, which are technically called the reduced levels, are obtained by the constant addition and subtraction of the numbers contained in the columns of "Rise," and "Fall," the former being considered as positive, and the latter as negative quantities; thus, assuming the level of the starting point A as the datum, we have the first forward station 5.83 feet higher than the datum, therefore in the column of reduced levels it is marked + (plus); next we have a fall or negative quantity of 6.90 feet, which must be subtracted; but as it is greater than 5.83, it shows that this station is below the datum line, by the difference between 5.83 and 6.90=1.07 feet, which is the depth of the second forward station *below* the datum line, and therefore is marked − (minus); the next is likewise a fall of 7.84, and as our last result was below the datum line, this additional negative quantity will take us still lower by its whole amount; it must, therefore, be added to 1.07, giving 8.91 feet for the depth of our third forward station below our datum, and it is therefore entered in the column of reduced levels with a minus sign. The next is also a fall of 9.78, which, applied as the last, gives 18.69 for the depth of the fourth forward station below the datum. The ground then rises again, and we have an ascent of 13.67 feet, which will bring us nearer to our datum ; and as it diminishes our depth below the datum line it must be subtracted from the last result; thus, 18.69 — 13.67 = 5.02 feet for the depth of the fifth forward station below the datum ; we have then a rise of 15.55, which will carry us above the datum by the amount of difference between it and 5.02, leaving 10.53 feet for the height of the sixth forward station above the datum line ; the next is a fall of 4.74, which diminishes our height by that quantity, and therefore must be subtracted from 10.53, leaving 5.79 as the height of the seventh forward station above the datum.

In like manner every other pair of sights in our example was reduced, applying each difference of the back and forward readings with their proper signs, until, at the close of the work, the point B (the last forward station) was found to be 49.33 feet above the datum line, or level of the starting point A.

The reduction of levels becomes a simpler operation when the height of the bench mark (used as a starting point) above the intended datum line is known ; thus (in our example), suppose the height of the bench mark A was 100 feet above the level of high-water Trinity mark at London Bridge, and that it was intended to assume the level of that mark as the datum line of our section ; then 5.83 feet, the rise to the first forward station, must be added to 100, giving 105.83 for the height of the ground at the point *a* above datum ; next from 105.83 subtract the fall 6.90, which gives 98.93 for the height of the point *b* above datum ; then from 98.93 subtract 7.84, which gives 91.09 for the height of *c* above datum ; and in like manner, by adding the quantities of rise, and subtracting those of the falls, the whole line of levels may be reduced to the line assumed as the datum.

As a proof of the accuracy of the arithmetical operation, the columns of back and fore sights should be added up, and the lesser sum subtracted from the former ; the result of the agreement with that by the reduced levels is a proof of accuracy. Likewise another proof may be obtained by adding up the contents of the column of rise and fall ; and if upon taking the lesser sum from the greater, the remainder represents the same quantity as obtained by both the other operations, there can be no doubt of the correctness of the reductions of the levels, as in our example. By the reduced levels, the height of B above A is 49.33 feet. The sum of the back readings is 87.95, and that of the forward readings 38.62 ; their difference also gives 49.33 for the height of B above A ; and, lastly, the sum of the rises is 54.88, and that of the falls is 5.55, the difference giving, as before, 49.33 feet.

It is, perhaps, to be recommended that the observer should reduce his levels as he proceeds in the field, as it will occupy but very little time, and can be frequently done while the staff-man is taking a new position; besides, the observer will frequently be able to detect by the eye if he is committing any glaring error, as, for instance, inserting a number in the column of rises, when it ought to occupy a place in that of the falls, the surface of the ground at once reminding him that he is going down hill instead of ascending.

If the foregoing method of reducing levels be found difficult or troublesome, on account of the introduction of plus and minus signs, they can be dispensed with as well as the columns of "Rise" and "Fall" by proceeding in the following manner. Assuming the starting point to be any even number of feet high; or, what is the same thing, assume a datum line any even number of feet below the starting point, as 100 or 1000, taking care that your choice falls upon a number greater than the number of the whole fall you are likely to experience in the operation; then from this assumed height *subtract* the reading of the forward staff, and to the remainder *add* the reading of the back staff; the result will be the height of the first forward station above the assumed datum line; then from this height subtract the next forward reading, and to the remainder add the reading of the back staff; the result will be the height of the second forward station above the assumed datum, and so on throughout the whole levelling operation. The difference between any two of the readings will be the difference of level between the corresponding points on the ground.

By way of illustration, we will reduce part of the foregoing example after this manner, and the student can adopt whichever method he may consider the best.

Back Sight.	Fore Sight.	Reduced Levels.	Remarks.
		100.00	Assumed datum.
13.71	7.88	7.88	
		92.12	
		13.71	
		105.83	Height of 1st forward station
9.40	16.30	16.30	above assumed datum.
		89.59	
		9.40	
		98.93	Height of 2d do. above do.
3.87	11.71	11.71	
		87.22	
		3.87	
		91.09	" 3d do. " do.
2.63	12.41	12.41	
		78.68	
		2.63	
		81.31	" 4th do " do.
14.62	0.95	0.95	
		80.36	
		14.62	
		94.98	" 5th do. " do.
17.00	1.45	1.45	
		93.53	
		17.00	
		110.53	" 6th do. " do.
10.66	15.40	15.40	
		95.13	
		10.66	
		105.79	" 7th do. " do.
2.87	17.00	17.00	
		88.79	
		2.87	
		91.66	" 8th do. " do.

The preceding will, we trust, be found sufficient to make ourselves understood upon the subject of reducing levels. If, after adopting the latter mode, it should be required to reduce them to the level of the starting point as a datum, nothing more is required than to take the difference between the height thus found and that of the assumed datum; thus, in our example, subtracting 100 (the assumed datum) from the height of the first forward station, gives 5.83 for its height above the starting point; next, from 100 subtract 98.93=1.07, making the second forward station that quantity below the level of the starting point, and so of the rest. But it may be done much easier after the section is made to the assumed datum, by drawing a line parallel thereto through the point A, or any other that may be determined on; thus the section may be at once adapted to any required datum line.

TO DRAW THE SECTION.

The levels being reduced, the surface line may be represented in the form of a section, as shown above the plan in Plate II. The vertical and horizontal scales of a section are seldom the same, which produces a caricatured representation; the vertical scale being so much greater than the horizontal, shows the depths of cutting and embankment required in the execution of road, railway, or canal works, with greater clearness than if both scales were equal. The plans and sections of projected works deposited with the Clerks of the

Peace of counties, and in the Private Bill office, to obtain the sanction of the Legislature, are mostly drawn to scales of four inches to one mile horizontal, and one hundred feet to one inch vertical; we have adopted these scales in our example, Plate II.

To make the section of our present example, first draw the horizontal line C D as the datum to which our levels were reduced, assume any point A as the starting point, then set off the measured distance from A to the first forward station a=519 links (see levelling field-book, page 50), at this point erect a perpendicular, and mark on it the height 5.83 of the first forward station, and connect the point A with this mark, and the result will show the surface line of the ground in that interval; next, from the same starting point A set off the point b, the second forward station, with the distance of 1315 links, as given in the levelling-book; but as this point is a minus quantity (see reduced level, page 50), that is, below the datum line, let fall a perpendicular, and set off on it 1.07 feet, which connect by a line with the former level, and the surface line from A to b will then be represented; then with the distance 1542 set off the point c, and on a perpendicular let fall therefrom, set off 8.91, which connect as before, and the section will be complete from A to c. In like manner, proceed with the rest of the reduced levels at the points d, e, f, &c., till the whole section is drawn.

Although, for the sake of clearness of description, we have desired the person plotting the section to draw the perpendicular, and thereon define the level point of the surface as he proceeds with setting off the horizontal distances step by step, yet in practice he will find it most expeditious in the first instance to place the chamfered edge of his ivory scale for the distances along the datum line, and at once to prick off the whole of the distances (or any convenient portion of them) successively as the numbers appear in the field-book ; then draw all the perpendiculars by means of a parallel ruler, or by a T square if the paper is properly fixed on a drawing table ; and, lastly, from the vertical scale prick off all the perpendiculars and connect those points, and the section will be made.

The distances given in the proper column of the field-book are supposed to be horizontal distances, and, in measuring them, care should be taken that they are as nearly such as possible (or they must afterwards be reduced thereto), otherwise the section will be longer than it ought to be. For the purpose of assisting the surveyor in making the necessary reduction from the hypothenusal to the horizontal measure, when laying down his section, we annex the following Table, showing the reduction to be made upon each chain's length, for the following quantities of rise, as shown by the reading of the staves :—

Rise in feet for one chain.	Reduction upon one chain in links and decimals.
1	0.01
2	0.04
3	0.11
4	0.19
5	0.29
6	0.44
7	0.56
8	0.74
9	0.94
10	1.16
11	1.40
12	1.76
13	2.01
14	2.24
15	2.61
16	2.99
17	3.39
18	3.76
19	4.23
20	4.64

The section can be referred to any other datum than the one by which it was produced ; as, for instance, let it be required to refer to section, Plate II., to a datum line 100 feet below the point A ; all that is requir- ed to be done is, to draw a line E F parallel to C D, at 100 feet below it ; then, by drawing perpendiculars from the surface line to this new datum, as shown by the dotted lines, the transfer will be complete, as the height of any point can be measured by the scale of the sec- tion. We need not go through a further explanation of this subject, as an inspection of our engraved exam- ple will explain whatever further may be required.

WORKING SECTION.

For the purposes of carrying into execution any work, the section should be much more minute than is requisite for general purposes ; it is then called a working section. The following are the field notes taken for such a section, the line having first been carefully set out and a stake driven into the ground at the extremity of each chain's length : these stakes were about 18 inches long and 2 inches square (and were furnished by a country wheelwright at the price of ten-pence per dozen) ; every tenth stake was circular, and somewhat larger, and had an iron ring round its top, and together with every fifth stake had their tops painted white, the more easily to identify them ; they were all numbered (or considered to be numbered) from one end of the line to the other. Plate III. shows the section of the ground and railway at the extreme end of the line where the numbers terminate at 1103 chains or 13¾ miles and 3 chains ; we would recommend the student to plot this section from the notes several times, and to various scales, that he may not only better understand the subject, but also for the sake of practice, it being an actual example from the working section of a line of railway now completed and opened to the public.

FIELD NOTES—WORKING SECTION.

Rise.	Back sight.	Fore sight.	Fall.	Distance.	Reduced Levels.	Remarks.
feet.	feet.	feet.	feet.	Links.	feet.	
					270.72	Brought forward (from last page of NOTES).
	4.47	4.53	0.06	103300	270.66	
	4.53	9.22	4.69	103400	265.97	
4.15	9.22	5.07		103500	270.12	
4.83	5.07	0.24		103600	274.95	
4.49	6.36	1.87		103700	279.44	
4.67	6.14	1.47		103800	284.11	
4.52	6.62	2.10		103900	288.63	Side of clapping post of field gate in occupation road.
	2.10	2.24	0.14	103916	288.49	
0.95	10.42	9.47			289.44	Lower hanging hook of gate.
	9.47	13.22	3.75	103944	285.69	Centre of occupation road.
0.07	13.22	13.15		103956	285.76	Edge of road.
4.40	13.15	8.75		103966	290.16	Top of bank.
4.27	8.75	4.48		103976	294.43	Do. do.
0.16	4.48	4.32		104000	294.59	
	2.44	8.84	6.40		288.19	B. M. south side of line.
6.01	8.84	2.83		104100	294.20	
	0.74	2.18	1.44	104200	292.76	
	2.18	5.35	3.17	104300	289.59	
	6.77	7.28	0.51	104400	289.08	
0.03	7.28	7.25		104490	289.11	Edge of ditch.
	7.25	8.36	1.11	104492	288.00	Bottom of ditch.
4.79	8.36	3.57		104500	292.79	Stump, top of bank.
0.62	3.37	2.75		104600	293.41	
1.32	2.75	1.43		104700	294.73	
	1.10	2.25	1.15	104800	293.58	
	2.25	8.86	6.63	104900	286.95	Enter alder plantation.
	5.65	9.53	3.88	104920	283.07	
	9.53	11.50	1.97	105000	281.10	
0.33	5.85	5.52		105021	281.43	
	5.52	12.01	6.49	105100	274.94	
	12.01	12.87	0.86	105148	274.08	
2.10	12.87	10.77		105190	276.18	
2.18	10.77	8.59		105200	278.36	Foot of bank, which rises perpendicularly 1 foot.
7.19	8.59	1.40		105300	285.55	
3.80	8.22	4.42		105400	289.35	
1.45	4.42	2.97		105500	290.80	

FIELD NOTES—Working Section.

Rise.	Back sight.	Fore sight.	Fall.	Distance.	Reduced Levels.	Remarks.
					290.80	Brought forward.
	2.97	3.39	0.42	105600	290.38	
	3.39	5.51	2.12	105700	288.26	
	5.51	7.67	2.16	105800	286.10	
	5.41	6.68	1.27	105827	284.83	Edge of ditch.
	6.68	8.56	1.88	105832	282.95	Bottom of ditch.
2.48	8.56	6.08	6.30	105837	285.43	Top of bank.
	6.08	12.38	4.34	105854	279.13	Foot of bank.
	12.38	16.72	0.62		274.79	
	2.04	2.66	3.82	105900	274.17	
	2.66	6.48	2.38	105940	270.35	Edge of ditch.
	6.48	8.86		105944	267.97	Bottom of ditch.
2.86	8.86	6.00	1.58	105952	270.83	Top of bank.
	6.00	7.58	3.16	105960	269.25	Foot of bank.
	7.58	10.74	4.91	106000	266.09	
	3.33	8.24	0.91	106095	261.18	Top of bank.
	8.24	9.15	4.19	106100	260.27	Stump, side of bank.
	9.15	13.34		106105	256.08	Bottom of ditch.
1.69	13.34	11.65	1.15	106110	257.77	Edge of ditch.
	11.65	12.80	0.51	106200	256.62	
	3.62	4.13		106300	256.11	
0.75	4.13	3.38		106349	256.86	Foot of bank.
2.88	3.38	0.50	3.85	106359	259.74	Top of bank.
	0.50	4.35		106368	255.89	Bottom of side drain.
0.27	4.35	4.08	0.23	106386	256.16	Centre of parish road.
	4.08	4.31		106405	255.93	Foot of bank.
3.74	4.31	0.57	2.45	106415	259.67	Top of bank.
	0.57	3.02	0.41		257.22	
	0.99	1.40	1.43	106430	256.81	Foot of bank.
	1.40	2.83	1.58	106500	255.38	
	2.83	4.41	0.07	106600	253.80	
	4.41	4.48		106700	253.73	
0.46	7.80	7.34		106800	554.19	(Crosses foot-path at 106831).
2.93	7.34	4.41		106900	257.12	
3.67	4.41	0.74		107000	260.79	
4.20	10.63	6 43		107100	264.99	
5.06	6.43	1.37		107200	270.05	
5.79	10.76	4.97		107300	275.84	
3.85	4.97	1.12	0.05	107400	279.69	
	5.42	5.47		107500	279.64	
0.91	5.47	4.56	0.44	107600	280.55	
	4.56	5.00	0.56	107637	280.11	Edge of ditch.
	5.00	5.56		107640	279.55	Bottom of ditch.

FIELD NOTES—WORKING SECTION.

Rise.	Back sight.	Fore sight.	Fall.	Distance.	Reduced Levels.	Remarks.
					279.55	Brought forward.
3.16	5 56	2.40	0.56	107647	282.71	Top of bank.
	2.40	2.96		107654	282.15	Foot of bank.
1.58	2.96	1.38		107700	283.73	
8.84	9.45	0.61		107800	292.57	
5.11	8.44	3.33		107853	297.68	Enter plantation.
2.91	3.33	0.42	1.50	107857	300.59	B. M. on timber stub.
	12.78	14.28		107882	299.09	
4.27	14.28	10 01		107900	303.36	
8.75	10.01	1.26		107947	312.11	
10.42	14.49	4.07		108000	322 53	
1.21	4.07	2.86	0.49	108008	323.74	
	2.86	3.35		108024	323.25	
2.98	3.35	0.37		108047	326.23	
15.33	16.35	1.02		108098	341.56	Top of bank, edge of plantation.
	1.02	1.32	0.30	108100	341.26	
2.74	8.66	5.92		108200	344.00	
0.78	5.92	5.14		108300	344.78	
	5.14	8.43	3.29	108400	341.49	
	1.05	4.50	3.45	108500	338.04	
	4.50	4.94	0.44	108520	337.60	
	4.94	6.83	1.89	108530	335.71	Edge of bank.
	6.83	12.54	5.71	108540	330.00	Foot of bank.
	12.54	16.82	4.28	108600	325.72	
	1.11	9.04	7.93	108700	317.79	
	1.18	9.09	7.91	108800	309.88	
	1.57	9.70	8.13	108900	301.75	
	1.28	9.58	8·30	109000	293.45	
	1.44	9.41	7.97	109100	285.48	
	1.34	9.14	7.80	109200	277.68	
	1.15	8.12	6.97	109300	270.71	
	3.04	4.43	1.39	109386	269.32	Edge of ditch.
	4.43	6.22	1.79	109390	267.53	Bottom of ditch.
1.06	6.22	5.16		109400	268.59	Stump, top of bank.
	5.16	11.10	5.94	109405	262 65	Foot of bank.
1.81	11.10	9.29		109500	264.46	
3.10	8.87	5.77		109600	267.56	At post and rail fence.
1.17	4.63	3.46		109700	268.73	Edge of slope.
	3.46	7.06	3.60	109800	265.13	
	1.96	4.60	2.64	109900	262.49	Foot of slope.
	4.60	4.60		110000	262.49	
0.93	4.60	3.67		110149	263.42	
	4.58	5.03	0.45	110377	262.97	Stump, end of curve.
0.63	5.03	4.40			263.60	On rails at end of curve.
	4 40	4.58	0.18		263.42	B. M. foot of post.
3.53	4.58	1.05			266.95	Top of said post.

In taking levels for a minute section where the observations must be very numerous, and consequently the back and fore sights not very far from each other, the observer will frequently be able to make a number of observations at each setting up of the level at one side of his line, so that his instrument may be about equally distant from his back and fore observations. Due attention to this will save much time and labor, and experience will enable the surveyor at a glance to see where he can set up his level at every remove forward with the greatest advantage. Upon looking down our field notes above, it will be seen it seldom occurred that only one back and one fore sight was obtained at a setting up of the level, and this only took place where the ground was very steep : by the first setting up of the instrument four forward sights were observed, and of course as many back ones ; thus the first back sight was 4.47, the corresponding fore sight 4.53 ; this latter number was also placed as the back sight for the next observation, which was 9.22 ; this number was in like manner placed as the back sight for the next forward observation, 5.07, which also became the back sight for the last forward observation we could obtain at that setting up of the instrument, namely, 0.24 : it should here be remarked that there was a necessity to place each forward reading as a back observation to the next forward reading, otherwise the difference of level between each point of observation would not have been obtained without more arithmetical work ; the numbers otherwise only show the difference of level between each and the first point of observation ; besides, by this arrangement, the whole section is continuous, however numerous the intermediate observations may be, and having the distances opposite, the whole can be plotted off with facility. The columns of Rise and Fall need no observation after what has already been said upon this subject. The column of distances denotes the continuous measurements from the commencement, Gunter's chain being the unit employed. Our notes commence at the 1033rd chain, and terminate with the end of the work, 1103 chains and 77 links, which we consider an ample extract for the purposes of the student. The column headed " Reduced Levels " contains the height of each point of observation above the datum line, which in this case was Trinity high-water mark, London Bridge : these numbers are obtained by adding the "rises" and subtracting the "falls" from the preceeding reduced level, which in our notes commence with 270.72 feet.

THE SECTION.—SEE PLATE III.

The datum line must be drawn, every chain should then be pricked off and the perpendiculars erected ; the chains or stakes should then be numbered beneath the datum line, to prevent mistake, and just above the datum line the height of the surface at each stake should also be inserted ; then the said heights can be pricked off upon the perpendiculars respectively, and the intermediate heights plotted from the field notes without fear of error, which otherwise, without great care, would be likely to occur in consequence of so many points falling near to each other, unless the scale be very large : the horizontal scale of the example is 1 inch to 5 chains, and the vertical scale 1 inch to 25 feet. Having drawn the undulating line of the surface through these points upon the perpendiculars, the gradients or intended line of railway may next be laid down ; the extreme left-hand point was given, being the level of the rails at the point of junction with another line. The railway is represented by two parallel lines, the upper one being the upper surface of the rails, and the lower one the bottom of the ballasting or formation level being 2.25 lower than the surface of the rails ; for a short distance the line is level, then it rises at the rate of 20 feet per mile, for the two-fold object of diminishing the great cutting and of getting sufficiently high over the road at stake 1064, to allow (with the lowering the surface of the said road a small quantity) of sufficient headway for the public carriages to pass under the railway ; from this point the line falls at the rate of 20 feet per mile for a considerable distance, the object being to get as low down as possible further to the eastward, where there was to be a considerable embankment, and by these means such embankment was reduced in dimensions ; and furthermore, the earth from the cutting to the right of the road was to be taken eastward to form the said embankment, and therefore the down-hill gradient was favourable for carrying on the work as well as for the drainage of the cutting. Part of the earth from the large cutting was also to be taken to the eastward; the ascending gradient, up to the bridge, was unfavourable for this purpose, however, so far as the bringing out the bottom of the cutting, the upper part being brought down by means of inclined planes; the ascending gradient was unavoidable in this case, but by judiciously working the excavation, little inconvenience and extra expense attended it. Each change of gradients is denoted by a strong vertical line from the datum to the point of change, and the height marked thereon. The quantity of earth-work to form the cuttings and embankments with different slopes should be written upon them, as shown in our example; also over the line of figures denoting the height of the surface above the datum should be placed the depth of the cutting from the surface to formation level at the same point, or the height of the embankment, as the case may be; these heights and depths are those from which the calculations of the quantities are to be made, and therefore must be strictly correct; they should not be taken from the section by the scale, but should be obtained by calculation; the former method being liable to error. The calculation may be thus performed. Let it be required to find the depth of cutting at stake No. 1083, where the height of the surface above datum is 344.78 feet; at stake No. 1064, the height of formation level above datum is 269.20, from which point the gradient descends at the rate of 20 feet per mile, or 0.25 feet per chain, towards No. 1083 ; the distance from 1064 to 1083 is 19 chains, which multiplied by 0.25, gives 4.75 for the fall of the railway in the interval between the two points ; consequently the height of the railway above datum at No. 10.83 is 269.20, minus 4.75 = 264.45 ; this sum, subtracted from the whole height of the surface, gives 344.78 — 264.45 = 80.33, for the depth of the cutting at that point, and so of all the remaining numbers. After giving the above particulars nothing need be added upon this subject. It may be worth observing, that in laying down the gradients care should be had so to dispose them as to produce the minimum quantity of work in the execution, and that the cuttings should equalize the embankments, or, if anything otherwise, they should be a little in excess, to allow for subsidence or slips in the embankments. The facilities for working the excavations and carrying the earth to bank should also be considered ; a down-hill gradient in that direction is most suitable, provided it can be obtained without interfering with other and often more important considerations ; the drainage of the works during the formation and after the line is completed should also be considered at the time of determining the gradients. We have inserted (Table I. at the end of the work) a very extensive and useful Table of Gradients, which is sufficiently self-explanatory as not to require further notice.

When a surveyor is required to level through a country in a perfectly straight line, and has not the advantage of its being picketed or poled out, his only means to keep a rectilineal course is by ascertaining, as accurately as possible, the magnetic bearing of one extremity from the other, and work in that direction by means of a compass. We once had business of this kind, and determined the bearing of our intended line from the map of the Ordnance survey (allowing for the variation of the needle), and after pursuing the route thus determined, we were surprised and delighted at finding how exactly we came to our required point, convincing us (if a proof had been required) how justly the public confidence has been placed in our national survey.

It is seldom the case in practice that the instrument can be placed precisely equi-distant from the back and forward staves, on account of the inequalities of the ground, &c. It would appear, therefore, to be necessary, to make our results perfectly correct, to apply to each observation the correction for curvature and refraction, as explained in the early pages of our book; this, however, we believe, is seldom done unless in particular cases, where the utmost possible accuracy is necessary, on account of the smallness of such correction, as may be seen by referring to our Table, page 13, where the correction for eleven chains is shown to amount to no more than $\frac{1}{100}$ of a foot; and as the difference in the distances of the instrument from the back and fore staves can in no case equal that sum, it is evident that such correction may be safely disregarded in practice.

Several machines have been constructed or designed for the purpose of describing a section of any ground passed over by the instrument, which at the same time would register the distance passed over, as well as the undulations: perhaps the best of this kind was the one designed and constructed by George Edwards, Esq., Civil Engineer, of Lowestoff, which is fully described and illustrated in the forty-fourth volume of the "Transactions of the Society of Arts," page 123, to which we refer. The use of such machines, however, must, from the nature of the work to be performed, be of a very limited character.

We have now described the leading principles and practice of levelling as employed in engineering operations; and although our observations may appear to be confined to its applicability to railroad purposes, yet the intelligent student will find no difficulty in applying to practice the same principles to every other branch of the profession where levelling operations may be required. We might indeed have multiplied instances and examples which would in reality have had no other effect than to swell our volume, as it must have been, to a great extent, but simply a repetition of the details already given.

Before closing this subject we cannot refrain from stating, that it has long been our opinion that if a register could be kept by some public body (as the Institution of Civil Engineers) of the height of particular spots throughout the kingdom, above some given datum, as Trinity high-water mark, London Bridge, or any other that might be agreed upon, such a record would be invaluable both in a particular and national point of view ; to the engineer and geologist it would be most important, and the whole register could be prepared from time to time at a trifling cost, if each engineer and surveyor would but contribute to the common stock by sending to head-quarters the level of any particular spots as he, in the course of his professional engagements, may have opportunity of determining them. We consider that no time is likely to be so favourable for the purpose as the present, as nearly the whole country has been levelled over for railway purposes within the last few years ; and no doubt the field notes of the greater part are still in existence from which a great many such standard levels could be extracted by the parties who took the levels, and which in a few years it will be impossible to eliminate. By way of showing more fully our meaning, we have extracted from our own levelling books a few such standard levels, and arranged them after the manner we have above alluded to.

COUNTY OF KENT.

Height in feet above Trinity high-water mark, London Bridge.

Upper edge of tablet over door of No. 1 Martello Tower, near Folkstone	256.4
Top of first milestone on the road from Folkstone turnpike to Dover	402.9
Top of second milestone, do	510.7
Surface of ground at Folkstone turnpike gate	534.6
Dock wall at Dover, opposite Railway Office	7.4

COUNTY OF SURREY.

Surface of ground at New Chapel Turnpike gate	188.0
Waste board of Godstone Ponds, back of White Hart Inn	319.2
Top of twentieth milestone (from Westminster Bridge) on the road from Godstone to East Grinstead	287.6
River Medway (tributary stream) meadows, west side of turnpike road, at Blundley Heath	151.2
Broadham Green, near Oxted, foot of pointing post	268.2

COUNTY OF SUSSEX.

Honey-pot Lane, South Chailley Common	122.5
Gullage Farm, source of the Medway, near the barn	324.6
Waste weir canal (east side of Lindfield)	82.1
Summit of South Downs at Plumpton Plains	682.5
" " at Mount Harry	573.3
Turnpike road, Brighton to Lewes, near the barracks	85 1
Cross roads, at Turner's Hill turnpike gate	535.2

LEVELLING WITH THE THEODOLITE.

The application of the theodolite to the practice of levelling is an operation of great simplicity. We must suppose the reader to be already acquainted with the construction and method of measuring angles with that valuable instrument ; and those who have no such knowledge, we refer to the Treatise on Mathematical Drawing Instruments spoken of, where every particular respecting it may be found. The ordinary 5-inch theodolite, of the best construction, is the one we recommend to the use of the surveyor, it being sufficiently accurate for most purposes that fall within his province, and is convenient to use on account of its portability. A larger theodolite is seldom employed, except on surveys of great extent upon trigonometrical principles, as those of the United Kingdom under the direction of the Board of Ordnance, where theodolites of 3 feet diameter have been employed to obtain the requisite degree of accuracy.

To use the theodolite in the common purposes of levelling, it is only necessary to set the instrument up at every spot on the line of country to be levelled, where the inclination changes, without regard to the minor inequalities of the surface, taking care that the adjustments have been carefully examined and rectified, as explained in the book above alluded to, especially those adjustments which set the line of collimation, and the spirit-level attached to the telescope, parallel to each other. Then set the instrument level by means of the parallel plate screws, and direct an assistant to go forward with a staff, having a vane, or cross piece, fixed to it exactly at the same height from the ground as the centre of the axis of the telescope. Having gone to the forward station, the assistant must hold the staff upright, whilst the observer measures the vertical angle, which an imaginary line connecting the instrument and staff makes with the horizon ; the instrument and staff should then change places, or, to save time, another staff should take the place of the instrument, and the instrument be removed to the former staff, and from thence the same angle should be taken back again, and the *mean* taken as the correct result.

The distance must then be measured, which will furnish all the data required to find the difference of level between the places of the instrument and staff ; this, it will appear evident, is a matter of trigonometrical calculation,* the measured distance being considered as the hypothenuse of a right-angled triangle, of which the perpendicular is the difference of level. It scarcely appears necessary to give the rule for the calculation, but for the sake of uniformity we shall do so.†

Add together the logarithm of the measured distance, and the log. sine of the observed angle; the sum, rejecting 10 *from the index, will be the log. of the difference of level, in feet or links, &c., the same as the distance was measured in.*

If the distance be measured with Gunter s chain, the result (in chains) can at once be obtained in feet, by

* Capt. Frome's Work, in 8vo, published 1840.

† See Appendix I.

simply adding to the above two logarithms the constant 1.8195439, which (10 being rejected from the index) will give the log. of the height in feet.

In this manner, by considering the surface of every principal undulation as the hypothenuse of a right-angled triangle, the operation of levelling may be carried on with great rapidity, but, it must be remarked, without pretensions to great accuracy ; in fact, in that particular, the use of the spirit-level will never be superseded.

Another method of applying a theodolite to the purposes of levelling was introduced by Sir John Macneill. He caused to be constructed, by Messrs. Troughton and Simms, a more powerful instrument for the purpose. It was a combination of the level and the theodolite. He set it up at the foot of an inclination, and sent a man on with a staff, as above described ; and whilst the observer was looking through the telescope, another assistant walked along the line, holding up another staff at every rise and hollow of the intervening surface, and thereby the observer could note how much such rises and hollows were below the line of his vision.

The distances from the instrument to the points where the staves were held up could then be measured, and the section drawn by simply ruling a line at the angle of elevation given by the instrument (or, more correctly, by computing the total elevation, and setting that up as a perpendicular, and drawing the hypothenusal line thereto), and marking thereon the measured distances, and from such marks drawing perpendiculars of the various lengths indicated by the staff at its different positions: a line connecting the extremities of the perpendicular will represent the section of the surface line.

Instead of measuring the distances, Sir John Macneill had attached to the eye-end of the telescope a beautifully made wire micrometer, similar to those applied to astronomical telescopes, by which he could tell with sufficient accuracy the distances required. This method of levelling, like the former by the theodolite, will give but a general approximation to the truth, depending in a great degree upon the quality of the instruments, and the care bestowed upon the operation.

PART III.

COMPUTATION OF EARTH-WORK—ROAD-MAKING—THE CLINOMETER, ETC.

WE have now to show the manner of applying a section to practical purposes. If the object to be attained is the making of a railroad, it is essential that it be formed as nearly level, and as perfectly straight, as the surface of the ground will admit of; for the nearer it approximates thereto, the more profitably will it be worked when completed, as locomotive steam-engines perform the most work with the least expense when the resistance they have to overcome is uniform and invariable. The same remarks hold with respect to a turnpike road; but the inclinations on the latter may be made greater and more variable, being worked by animal power, which is capable of putting forth, on a sudden emergency, a greater exertion for a short time, which is not the case with elemental or mechanical power beyond limits much short of what an animal is capable of.

Sir Henry Parnell, in his valuable Treatise on Roads, recommends that a road should not be made steeper than 1 in 35; that is, for every 35 feet in length of road surface, the difference of level will be 1 foot, that being an inclination which presents no difficulty to fast driving either in ascending or descending. But on a line of railroad to be traversed by locomotive engines, no rate of inclination, or gradient, as it is called,* should exceed 20 feet in a mile, or 1 in 264. To draw the lines of proposed surface (or gradients) upon a section, which shall be the most suitable for the purposes intended, and at the same time to be the most economical in the execution, that is to say, to have the least possible quantity of earthwork in cuttings and embankments, requires judgment and experience ; no definite rules can be given for this purpose, as no two sections present the same undulating surface. There is one material point we would suggest, and which should be carefully attended to ; viz., that for every piece of cutting, there should be an equal, or *rather less*, quantity of embankment. We say rather less because every newly formed embankment experiences a settlement to a greater or less degree, and therefore more earth will be required to raise it to a proper level. The excess of the cuttings above the embankments should never be great, otherwise the surplus would have to be disposed of in mounds, termed spoil-banks. In no case whatever should the required embankments exceed in cubical contents the quantity of cuttings ; for then

---

* Sir John Macneill in his preface to his valuable translation of M. Navier's little work on the "Means of comparing the respective Advantages of different Lines of Railway," says, "I have rendered the word *peute* by *slope* in preference to *inclination*, *inclined plane*, or *gradient*, considering the two former, though generally used, as improper expressions ; and the latter, to say the least of it, as having so little to recommend it, that I hope it will have an extremely short existence in our nomenclature."

a serious difficulty occurs—land has to be purchased for the purpose of digging earth to supply the deficiency, which is usually called side cutting.

Suppose in the cut below the upper figure to represent the section of an old line of road, and that it were required, by cutting and embankment, to reduce it from its present hilly surface to one uniform rate of inclination from the point A to the point B. The lower

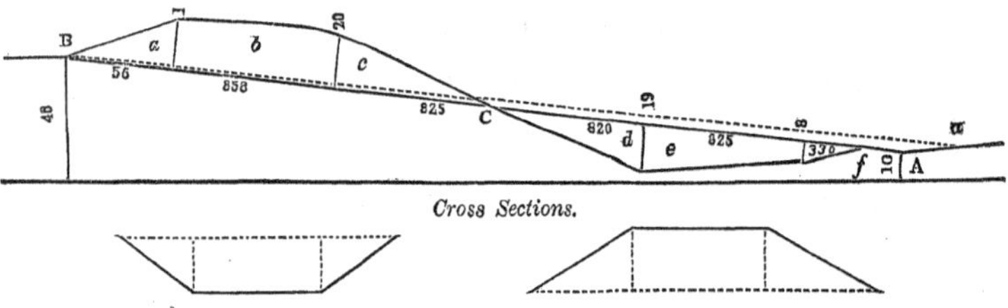

Cross Sections.

extremity A is 10 feet above the datum line of the section, and the higher point B 46 feet above the datum; consequently, $46 - 10 = 36$ feet, the rise from A to B, and the distance 4356 feet, which, divided by the rise (36), will give 1 in 121 for the rate of inclination the road may be brought to.

Upon the section draw the straight line A B, which will show the extent of cutting and embanking to be made. The number of cubic yards of earth to be removed in the cutting between the points B and C, and the cubical contents, in yards, of embankment between C and A, may then be computed in the following manner:

Divide the quantities of cuttings and embankments as shown upon the longitudinal section, into triangles and trapeziums, determined by the undulations of the surface lines, as shown in the above engraving, where, in the cuttings, *a* and *c* are triangles, *b* a trapezium; and in the embankments *d* and *f* are triangles, *e* a trapezium. The form of the excavation and embankment is shown by the transverse or cross sections. Let the width of the roadway (or base of the cutting, and top of the embankment) be 50 feet, including the footpath, &c., on each side; the slope of the cutting to be 1½ to 1, that is, 1½ horizontal to 1 perpendicular; consequently, where the depth is 20 feet, the width of the slope at the surface will be 30 feet; the slope of the embankment to be 2 to 1, that is, for 19 feet perpendicular, the base is to be 38 feet. With these data, the cubical quantities, as computed by the valuable Tables of Sir John Macneill,* are as follows:

Excavation........	81517 yards.
Embankment......	57081 "
	24436 surplus cutting.

We have an excess of 24436 cubic yards of excavation, which is a quantity far too great. In order, therefore, to make the quantity of cutting and embankment more nearly balance each other, it would be necessary to continue the embankment beyond the point A, which would lengthen the inclination, as shown by the dotted

* "Tables for calculating the cubic quantity of earthwork in the cuttings and embankments of canals, railways, and turnpike roads." By Sir John Macneill, Civil Engineer, F.R.A.S., &c.

line drawn from the point B to a; this dotted line would now represent the proposed surface of the road. By such means we diminish the quantity of cutting, and, at the same time, increase that of the embankments; and also by lengthening the inclination, we reduce its steepness. The alteration of the proposed surface line must be so made, that the cubical quantities of excavation and embankment are nearly equal; leaving, however, a preponderance in favour of the latter of about 10 per cent. to supply the deficiency occasioned by the consolidation and shrinking of the earth; and if any portion of the excess be then remaining, it may be disposed of in flattening the slopes of the embankments, when no more convenient mode presents itself.

The quantities of earthwork on a given section depend upon the arrangement and disposition of the *gradients*, or proposed surface lines; and there is no practical consideration of more consequence to the engineer, in laying out a proposed line of surface upon a section, especially if it be of any great extent (as the present projected lines of railway), than the most judicious distribution of the cuttings and embankments, which should not only be nearly equal to each other in quantity, but the circumstances must be considered under which the various embankments have to be supplied, it not being alone sufficient that for every hollow on the section there should be a corresponding protuberance, but that such protuberances be advantageously situated for filling the hollows ; for otherwise the work assumes a character of difficulty, in consequence of the great additional expense of removing the earth to a considerable distance ; and if, in addition, the material has to be conveyed up an ascent, it will be more tedious in the execution.

Knowing the value of practical examples in elementary books, we shall here give the calculations of the above results in full, both by the common method ; viz., *The Prismoidal Formula*, and Sir John Macneill's Tables, by which the saving of labour by the use of the Tables will be made apparent.

Prismoidal Formula.—The area of each end added to four times the middle area, and the sum multiplied by the length divided by 6, will give the solid content. If the measures used in the calculation are yards, the result will be the content in cubic yards ; but if they are feet, the result must be divided by 27, to obtain the corresponding number of yards.

CALCULATION OF THE TRIANGULAR PORTION *a*.

Height 0.

		2) 18	
Height........	18	9	mean height.
Slope	1.5	1.5	slope.
	9.0	4.5	
	18	9	
	27.0	13.5	
Base.........	50	50.0	base (bottom of cutting, or top of embankment).
	77	63.5	

Brought forward,	77	63.5	
Height........	18	9	mean height.
	———	———	
	616	571.5	middle area.
	77	4	
	———	———	
Area of greater end }	1386	2286.0	4 times middle area.
		1386.0	area of greater end.
	———	———	
		3672	
		561	length.
		———	
		3672	
		22032	
		18360	
		———	
		6) 2059992	
		———	
		3) 343332	cub. content in feet.
		———	
		9) 11444	
		———	
		12716	cub. content in yards.

COMPUTATION OF *b*.

Height 18. Area, as before, 1386.

20 height.		18 }	heights.
1.5 slope.		20 }	
——		—	
10.0		2) 38	
20		—	
——		19	mean height.
30.0		1.5	slope.
50 base.		——	
——		9.5	
80		19	
20 height.		——	
——		28.5	
1600	{ area between *b* and *c*.	50	base.
——		——	
		78.5	

```
Brought forward,      78.5
                      19       mean height.
                     ------
                     7065
                     785
                     ------
                    1491.5     middle area.
                        4
                     ------
                    5966.0   4 times middle area.
                    1386     area of lesser end.
                    1600     area of greater end.
                     ------
                    8952
                     858     length.
                     ------
                    71616
                   44760
                  71616
                 ---------
              6) 7680816
                 ---------
              3) 1280136     cub. content in feet.
                 ---------
              9) 426712
                 ---------
                  47412      cub. content in yards.
```

COMPUTATION OF *c*.

Area 1600, as before.

```
      20 }
       0 } heights.
      --
   2) 20
      --
      10   mean height.
       1.5 slope.
      ---
      50
```

Brought forward,

```
        50
        10
       ----
       15.0
        50   base.
       ----
        65
        10   mean height.
       ----
       650   middle area.
         4
      -----
      2600   4 times middle area.
      1600   area of greater end.
      -----
      4200
       825   length.
     ------
     21000
     8400
    33600
   -------
 6) 3465000
   -------
  3) 577500  cub. content in feet.
     ------
 9) 192500
    ------
     21389   cub. content in yards.
     ------
```

$a =$ cub. content..................12716
$b =$ cub. content..................47412
$c =$ cub. content..................21389

Total cuttings...................81517 cub. yards.

COMPUTATION OF EMBANKMENT *d*.

```
 19 height.             0 }
  2 slope.             19 } heights.
 ---                  ---
 38                2) 19
 50 base.             ---
 ---                  9.5   mean height.
 88                   2     slope.
 19 height.          ----
----                 19.0
792                  50      base.
 88                  ----
----                 69
1672 area             9.5   mean height.
----                 ----
                     34.5
                    621
                    -----
                    655.5    middle area.
                      4
                   ------
                   2622.0   4 times middle area.
                   1672     area of greater end.
                   ----
                    4294
                     820    length.
                   ------
                   85880
                  34352
                 --------
              6) 3521080
                 --------
              3) 586847      cont. in cub. feet.
                 ------
              9) 195616
                 ------
                  21735      cont. in cubic yards.
```

COMPUTATION OF e.

Height 8. Area, as before, 1672.

8 height.	19 }	heights.
2 slope.	8 }	
16	2) 27	
50 base.		
	13.5	mean height.
66	2	slope.
8 height.		
	27.0	
528 area.	50	base.
	77	
	13.5	mean height.
	385	
	231	
	77	
	1039.5	middle area.
	4	
	4158.0	4 times middle area.
	1672	area of greater end.
	528	area of lesser end.
	6358	
	825	length.
	31790	
	12716	
	50864	
	6) 5245350	
	3) 874225	cub. content in feet.
	9) 291408	
	32379	cub. content in yards.

COMPUTATION OF f.

Area, as before, 528.

```
  8 }
  0 } heights.
  —
2) 8
  —
  4   mean height.
  2   slope.
  —
  8
 50   base.
 —
 58
  4   mean height.
 ———
232   middle area.
  4
———
928   4 times middle area.
528   area of greater end.
————
1456
 330  length.
—————
43680
4368
——————
6) 480480
——————
3) 80080   cub. content in feet.
——————
9) 26693
——————
    2966   cub. content in yards.
```

d = cub. content..............	21735
e = cub. content..............	32379
f = cub. content..............	2966
Total embankment............	57080 cubic yards.

The same quantities computed by Sir John Macneill's Tables.

THE CUTTINGS.

COMPUTATION OF *a*.		COMPUTATION OF *b*.	
Tabular No........	= 22.67	Tabular No.........	= 55.26
Length...........	561	Length............	858
	2267		44208
	13602		27630
	11335		44208
Cont. of *a* in cub. yards	=12717	Cont. of *b* in cub. yards	=47413.08

COMPUTATION OF *c*.

Tabular No..................	= 25.92
Length.....................	825
	12960
	5184
	20736
Cont. of *c* in cub. yards....	= 21384.00

THE EMBANKMENTS.

COMPUTATION OF *d*.		COMPUTATION OF *e*.	
Tabular No......	= .3519	Tabular No.......	= 5000
Base............	50	Base	50
	17.5950		25.0000
Tabular No.....	+ 8.914	Tabular No.....	+ 14.247
	26.509		39.247
Length........	820	Length.........	825
	530180		196235
	212072		78494
			313976
Cub. content..	= 21737.380	Cub. content..	= 32378.775

COMPUTATION OF f.

Tabular No................	= .1481
Base......................	50
	7.4050
Tabular No................	+ 1.580
	8·985
Length....................	330
	269550
	26955
Cub. content..............	2965.050

RESULTS BY THE TABLES.

CUTTINGS.	EMBANKMENTS.
a = 12717.9	d = 21737.4
b = 47413.1	e = 32378.8
c = 21384·0	f = 2965.0
81515·0	57081.2

By comparing these results with those obtained by the former process, it will be seen that the cubical quantity of cuttings differs but two yards, and that of the embankments but one yard. The computation by the Tables may be abbreviated by using but one place of decimals, which would be sufficiently accurate for practical purposes. Our object is to show the calculations, by the Tables, in their greatest extent, which even then produce a great saving of labour, and, of course, a much greater probability of accuracy, in consequence of the fewer figures employed, than the former process.

It will be seen that the calculation of the embankments by the Tables is a longer process than that of the cuttings, the latter being done by simply multiplying a number taken from the Tables (answering to the height or depth at each end) by the length ; whilst, for the embankments, the tabular number is first multiplied by the base (or width of roadway), and to the product is added a second tabular number taken out at the same time as the first. The first series of Sir John Macneill's Tables contain the numbers corresponding to a base of 50, and a slope of 1½ to 1 (which is the slope of the cuttings in our example). But for a slope of 2 to 1, reference must be had to the second series of the same tables, which are applicable to every width of base, and from slopes varying from ½ to 1, to 3 to 1. We have adopted this example to show the calculations both by the *particular* and *general* Tables, as the first and second series of the valuable work referred to may be called.

The following is an extract from Sir John Macneill's preface to his Tables :—" All practical engineers are well aware, by experience, of the inconveniences which arise from the length of time necessary for calculating the cubic quantity of earthwork in the cuttings and embankments of canals, railways, and turnpike roads, especially when the section is of considerable extent, and the ground very uneven. As calculations of this kind are frequently, on a short notice, required to be completed within a limited period, the consequence is, that errors are almost sure to be made, as a multiplicity of figures is necessary, though the calculations in themselves are so very simple.

"To save time in making these calculations, and ensure accuracy in the results, were the principal objects I had in view in constructing the following Tables ; how far I have succeeded, must be left to the decision of practical men, for whose use they were intended, and who are best able to judge of their utility.

"An advantage may rise from the use of these Tables, which I had not at first contemplated. By the common but erroneous method of calculation, the cuttings may appear to be equal to the embankments ; yet when the work is carried into effect, a large quantity of earth may be required to make up the embankments, or there may be too much earth in the cuttings for the embankments, according to the shape or figure of the section, as will be shown hereafter. Such a circumstance as this cannot take place if the following Tables be used to ascertain the cubic quantities ; for, as they are calculated from the prismoidal formula, they will give the true cubic quantity in any cutting or embankment; and consequently, if the cuttings be laid down on the section to balance the embankments, they will be found in practice to do so, when the work comes to be executed.

"Contractors very frequently find that they have more earth to move than they had previously calculated upon from the section, and are, therefore, often great losers. This, in most cases, arises from erroneous calculations; for the common practice is, either to add the two extreme heights together, and to take half the sum for a mean height; or to take half the sum of the areas at each end for a mean area. Both these methods are erroneous; one makes the quantity too much—the other too little."

SLOPES, ETC.

As connected with the subject of earth-work, we may insert in this place some particulars respecting the arrangement of slopes in cuttings and embankments. They are usually expressed in terms of the height or depth of cutting, as half to one, one to one, two to one, &c., signifying that for every foot perpendicular, the cutting shall batter half a foot, one foot, two feet, &c.

The slope adopted must depend upon the nature of the material worked upon. Solid rock may be left perpendicular, whilst loose friable material, or sand, will stand but a very small angle with the horizon. The true criterion to judge of the proper slope to work to, is to observe, if convenient, what slope or angle the materials naturally assume when left to themselves. To determine this by measurement would be troublesome and tedious; but by the aid of a small instrument called a clinometer, the angle which any sloping surface makes with the horizon may be at once measured, and the ratio of the slope to the perpendicular, as one to one, &c., be readily deduced. As this very useful portable instrument is not generally known, we shall subjoin an engraving and description of it.

The following figure represents a clinometer, or, as it is called in some parts of the country, a *batter level.* It consists of a quadrant A B, of about two inches radius, attached to a flat bar C D, six inches long. The quadrant is graduated to degrees, from B towards A, and adjoining the divisions may be inserted, if required,

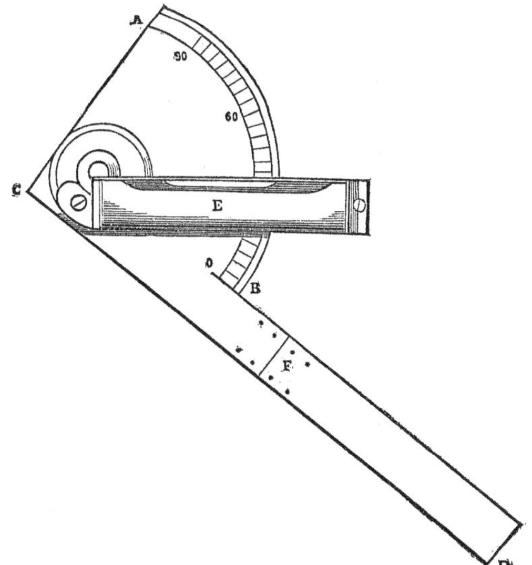

the corresponding ratio of the slopes, one to one, &c. An index bar, E, turns upon the centre of the quadrant, and carries a spirit-level by which the index may be set truly horizontal by the hand; and whatever angle is there denoted on the quadrant, will be that of the slope required. At F is a hinge-joint, by which the bar C D may be folded up, and the instrument can then be deposited in a box of very small dimensions, and carried in the pocket without inconvenience.

To use this instrument, open the hinge-joint, and rest the edge of the bar C D on the face of the slope to be measured; then gently move the index E round its centre until the attached spirit-bubble assumes a central position in its glass tube, and the angle, indicated by the index on the graduated arc, will at once measure the inclination. The ratio of the slope to the perpendicular is represented by the natural co-tangent of the angle thus measured; but as the observer may not have at hand a Table of natural co-tangents, &c., we have annexed a Table at once showing the slopes corresponding to the various angles of inclination likely to be required.

It will appear evident, that the longer the bar C D is, the more accurate will the measure of the slope be; but there is no necessity for the instrument to be encumbered with a long bar, which would destroy its portability, because it can easily be attached, by tying, to the end of a long straight rod, which can be furnished by any neighbouring carpenter, and the real slope of an undulating inclined surface can then be accurately measured.

Table of Slopes.

Slope or Batter to 1 foot perpendicular.		Ratio of Slope to perpendicular.	Angle of slope.				Slope or Batter to 1 foot perpendicular.		Ratio of Slope to perpendicular.	Angle of Slope.			
			With vertical.		With horizon.					With vertical.		With horizon.	
ft.	in.		°	′	°	′	ft.	in.		°	′	°	′
0	$\frac{1}{4}$	$\frac{1}{48}$ to 1	1	12	88	48	1	0	1 to 1	45	0	45	0
0	$\frac{1}{2}$	$\frac{1}{24}$ " 1	2	23	87	37	1	3	1$\frac{1}{4}$ " 1	51	20	38	40
0	$\frac{3}{4}$	$\frac{1}{16}$ " 1	3	35	86	25	1	6	1$\frac{1}{2}$ " 1	56	19	33	41
0	1	$\frac{1}{12}$ " 1	4	46	85	14	1	9	1$\frac{3}{4}$ " 1	60	15	29	45
0	1$\frac{1}{4}$	$\frac{1}{10}$ nearly 1	5	57	84	3	2	0	2 " 1	63	26	26	34
0	1$\frac{1}{2}$	$\frac{1}{8}$ to 1	7	8	82	52	2	3	2$\frac{1}{4}$ " 1	66	2	23	58
0	1$\frac{3}{4}$	$\frac{1}{7}$ nearly 1	8	18	81	42	2	6	2$\frac{1}{2}$ " 1	68	12	21	48
0	2	$\frac{1}{6}$ to 1	9	28	80	32	2	9	2$\frac{3}{4}$ " 1	70	1	19	59
0	2$\frac{1}{2}$	$\frac{1}{5}$ nearly 1	11	46	78	14	3	0	3 " 1	71	34	18	26
0	3	$\frac{1}{4}$ to 1	14	2	75	58	3	6	3$\frac{1}{2}$ " 1	74	3	15	57
0	3$\frac{1}{2}$	$\frac{7}{24}$ " 1	16	16	73	44	4	0	4 " 1	75	58	14	2
0	4	$\frac{1}{3}$ " 1	18	26	71	34	4	6	4$\frac{1}{2}$ " 1	77	28	12	32
0	4$\frac{1}{2}$	$\frac{6}{16}$ " 1	20	34	69	26	5	0	5 " 1	78	41	11	19
0	5	$\frac{5}{12}$ " 1	22	37	67	23	5	6	5$\frac{1}{2}$ " 1	79	42	10	18
0	5$\frac{1}{2}$	$\frac{6}{13}$ " 1	24	37	65	23	6	0	6 " 1	80	32	9	28
0	6	$\frac{1}{2}$ " 1	26	34	63	26	7	0	7 " 1	81	52	8	8
0	7	$\frac{7}{12}$ " 1	30	15	59	45	8	0	8 " 1	82	53	7	7
0	8	$\frac{2}{3}$ " 1	33	41	56	19	9	0	9 " 1	83	39	6	21
0	9	$\frac{3}{4}$ " 1	36	52	53	8	10	0	10 " 1	84	17	5	43
0	10	$\frac{10}{12}$ " 1	39	48	50	12	11	0	11 " 1	84	48	5	12
0	11	$\frac{11}{12}$ " 1	42	31	47	29	12	0	12 " 1	85	14	4	46

It is very important, in fixing upon the slopes for the sides of an excavation or embankment, to approximate very nearly to the inclination at which the ground would naturally stand without slipping ; for if they be made greater than necessary, a large quantity of labour, and of the surface of the ground, will be uselessly devoted. The proper slope for each particular soil can only be determined by observation and experience. " An embankment that would stand perfectly firm, and bear the action of the weather, when formed of sand, gravel, or the debris of rocks, and other materials that do not retain water in their fissures, would not last one winter, if it chiefly consisted of clay. The same remark applies with equal force to cutting, where it is made through a stratum of clay."* A slope of 1 to 1, that is, a slope of 45°, is found sufficient for ordinary earth; for clay 1½ to 1, or a slope of 33° 41′ with the horizon, may often be required, unless it can be mixed with open materials to prevent water collecting in the fissures produced by its shrinkage in dry weather. In other cases, so steep a face may be left as ¾ to 1, or even ½ to 1; and the slope that will be likely to stand may easily be judged of, by knowing the nature of the strata which will be cut through, and examining its state when exposed in the surrounding district."

At Boughton Hill, near Canterbury, there is a large cutting through London clay, which, together with the embankment at the foot of the hill, formed of the same material, has been constantly giving way. The slopes of the embankment have been flattened from time to time, and now assume some appearance of consolidation; but the slopes of the cutting near the summit of the hill continue to slip down upon the roadway. From some cross-sections we were able to take a short time since, it appears that the original slope of the cuttings was about 2 to 1, forming an angle with the horizon of 26° 34′; but the natural slope assumed by the soft clay where it has slipped is about 9°, or a little more than 6¼ to 1.

---

* Tredgold on Railroads, first edition, page 117.

ON SELECTING A LINE OF COUNTRY FOR A ROAD OR RAILWAY.

The choice of a suitable line of country for the formation of a turnpike-road, a railroad, or a canal, preparatory to the levels being taken, requires both judgment and care ; otherwise a fruitless expenditure of time in taking a number of trial-sections may be the result, if attended with no more serious and permanent inconvenience. A person undertaking such a work should previously devote a little time to obtain a knowledge of the country, its localities, its structure, and geological character : such knowledge will lead to the choice of several lines of direction, which appear to the eye as equally favourable ; it then becomes necessary to make such preliminary surveys as will enable the engineer to adopt the one which, under all circumstances, is likely to prove the most advantageous.

At p. 113 (1st edition) of the late Mr. Tredgold's work upon Railroads, we find the following observations upon this subject: "In order to facilitate the choice of a line as it regards the surface of the country, the engineer may be reminded, that even in the disposal which nature has made of hills and valleys there is much system. Those things which to the first glance of the better-informed, and at all times to the ignorant, appear to be without order or arrangement, are the result of the uniform action of natural causes, and are, in reality, capable of being traced and described with less difficulty than would be expected. Where a considerable tract of country is to be surveyed, the best index to its elevations and depressions is its streams and rivers ; these indicate every change of inclination, and, to the experienced eye, with considerable precision. It will also be observed, that each river has its system of valleys ; and except in a few instances, where the draining is effected by the outburst of an open stratum, a district, whose bounding ridge is easily traced, is drained by its river and system of valleys.

"Having formed a tolerable idea of the best direction for the road, the next step must be to make a more particular survey, with a view to fix nearly the precise line. We would recommend the principal engineer to have this done by rectangular lines, as infinitely superior to surveying by triangles, in giving him an exact knowledge of the surface of the country. Perhaps, with the assistance of a diagram, we shall be able to render the advantage of this method obvious.

"Let A B be a portion of the intended line, and C D the breadth of the country to be included in the survey. At any suitable distances choose stations, *a*, *a*, *a*, their distances apart depending on the changes of level, and let the principal line A B, and also the cross lines *b b*, *b b*, &c., be accurately levelled, and then drawn, as shown in the figure, on the plan of the line of road. If the distance *b b* is required to be considerable, perhaps an additional line in the principal direction may be necessary. The etched lines show the form of the surface at the lines A B, *b b*, *b b*, &c., on the plan ; and

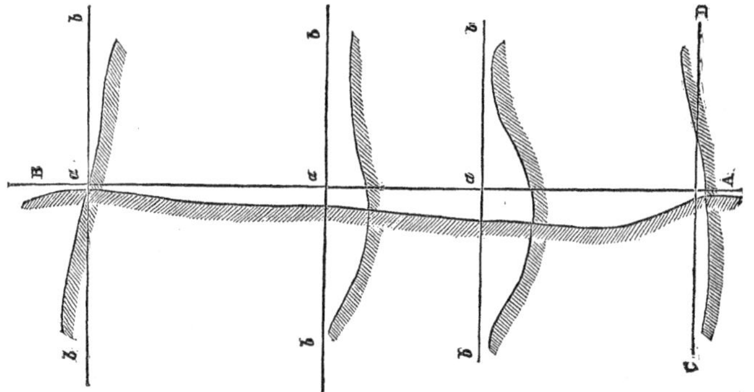

the latter being sections at right angles to A B, there is no difficulty in seeing the extent of cutting, or of embankment, that may be avoided by varying the position of the principal line. In fact, a plan of this kind, to a person familiar with sections, is better than a model of the country."

The most advantageous direction for a line, either of roadway or railroad, intended to connect two places, is evidently that of a right line, both horizontally and vertically : if one extremity of the line is more elevated than another, the straight line connecting them will be an inclined plane, having one uniform rate of inclination ; but if a uniform slope cannot be obtained in the direct line, it is necessary to deviate therefrom to obtain, as nearly as the circumstances of the country will admit, such an inclined plane, or at least to obtain continued progressive rises, avoiding as much as possible the introduction of useless ascents, that is, ascending where we must descend again, and *vice versâ*. When a line of road is encumbered with numerous and extensive useless ascents, the wasteful expenditure of power in the conveyance of goods is very great, as the number of feet actually ascended is increased many times more than is necessary, if each height, when once gained, were not lost again.

Sir Henry Parnell, in his valuable treatise on Roads, gives the following instances of this kind of roadmaking:—"As one instance, amongst others, of the serious injury which the public sustain by this system of roadmaking, the road between London and Barnet may be mentioned, on which the total number of perpendicular feet that a horse must now ascend is upwards of 1300, although Barnet is only 500 feet higher than London; and in going from Barnet to London, a horse must ascend 800 feet, although London is 500 feet lower than Barnet."

Another instance of this defect in road-engineering is observable in the line of the old road across the island of Anglesea, on which a horse was obliged to ascend and descend 1283 perpendicular feet more than was found necessary by Mr. Telford, when he laid out the present new line, as shown by the annexed Table:

	Height of summit above high water.	Total rise and fall.	Length.	
			Miles.	Yards.
Old road...	339	3540	24	428
New road...	193	2257	21	1596
Difference..	146	1283	2	592

In choosing the best direction for a line of roadway, the rate of inclination which can be obtained, with a moderate outlay in cuttings and embankments, is a consideration of greater importance than the mere maintaining of a direct line. For though the measured length of a circuitous route may be considerably greater than the length of a direct line, yet if the inclinations in the former case are much more favorable than those in the latter, it must be evident that more may be gained in speed, with the same expenditure of power, than is lost by the increase of distance. Thus, if two roads rise, one at the rate of 1 in 15, and the other at the rate of 1 in 35, the same expenditure of power will move a weight through 15 feet of the one and 35 feet of the other, at the same rate.

Upon the subject of the maintenance of turnpike roads, we shall annex an abstract of the General Rules for Constructing and Repairing Roads, laid down by the late Mr. Telford, and which is so fully treated upon in the important work of Sir H. Parnell on Roads.

SHAPE, OR TRANSVERSE SECTION.

The roadway should be 30 feet broad; the centre should be 6 inches higher than the level of the sides, where the junction of the surface, with the sloping edge of the foothpaths, or other defining bounds of the roadway, form the side channels; at 4 feet from the centre (on each side) the surface should be half an inch lower; at 9 feet, it should be two inches lower; and at 15 feet, its extreme edge, it should be 6 inches lower; this will give the form of a flat ellipse, which is well adapted for carrying off the water to the side channels, without making the cross section of the road too round, and allow the sun and wind to have a greater effect in evaporation, and keeping the road dry. In giving the surface one uniform curvature from side to side, the surveyor should use such a level as is described at page 111.

The footpaths should be 6 feet broad, and have an inclined surface of 1 inch in a yard towards the road; its surface should not be lower than the level of the centre of the road, and the edge should be sloped down (and covered with green sod) to meet the roadway, and form the side channel to carry off the water from the surface.

DRAINAGE.

All open main drains should be cut on the field side of the road fences, and should lead to the natural watercourses of the country ; in general they should be 3 feet deep below the bed of the road, 1 foot wide at bottom, and from 3 to 4 feet wide at top. Stone drains and culverts should also be made under the road, and continued to the open side drains, or ditches ; side channels (before named) must be made on the road side, with openings of masonry into the cross drains, to prevent any water lying on the road, it being necessary, in order to preserve the surface of a road perfect, that it be kept completely dry. All land springs ought to be carried from the site of the road by under-draining.

FENCES.

" All road fences should be kept as low as possible, never being allowed to exceed 5 feet in height, in order that they may not intercept the sun and wind, and diminish their effects in producing evaporation ;" and for the same reason no trees should be allowed to grow by the side of a road ; for by keeping the roads wet, they occasion the rapid wear of the materials of which they are formed.

ROAD MATERIALS.

The hardest description of stone should always be preferred, such as basalt, granite, quartz, &c. " The whinstones found in different parts of the United Kingdom, Guernsey granite, Mountsorrel and Hartshill stone of Leicestershire, and the pebbles of Shropshire,

Staffordshire, and Warwickshire, are among the best of the stones now commonly in use. The schistus rocks, being of a slaty and argillaceous structure, will make smooth roads, but they are rapidly destroyed when wet by the pressure of the wheels, and occasion great expense in scraping, and the constantly laying on of new coatings. Limestone is defective in the same respect. Sandstone is generally much too weak for the surface of a road ; it will never make a hard one. The hardest flints are nearly as good as the best limestone ; but the softer kinds are quickly crushed by the wheels of carriages, and make heavy and dirty roads. Gravel when it consists of pebbles of the harder sorts of stones will make a good road ; but when it consists of limestone, sandstone, flint, and other weak stones, it will not ; for it wears so rapidly, that the crust of a road made with it always consists of a large portion of the earthy matter to which it is reduced, and prevents the gravel from becoming consolidated, and the road from attaining that perfect hardness it ought to possess."* When the materials are stone, they should be broken to a size of a cubical form not exceeding 2½ inches in their largest dimensions, and should be capable of passing through a ring of that diameter. When it consists of gravel, the pebbles which are from 1 to 1½ inch in size only should be used for the middle part of the road ; all larger pebbles should be broken ; the small-

---

* Abridged from Sir H. Parnell on Roads, page 271.

er stones may be used for the sides of the roads and the foothpaths.

THE FOUNDATION AND DISPOSITION OF MATERIALS.

Before the foundation is laid, the surface on which it is to rest must be prepared, by making it level from side to side, and, if necessary, raising it so that the finished surface of the road may not be below the level of the adjoining fields. If the subsoil be wet and elastic, it must be rendered non-elastic by whatever means is best adapted to overcome the cause, as drainage, &c. The foundation should consist of a rough close-set pavement, of any kind of stones that can be most readily procured; those set in the middle of the road should be 7 inches in depth; at 9 feet from the centre, 5 inches; at 12 feet from the centre, 4 inches; and at 15 feet, 3 inches. They should be set with their broadest faces downwards, and lengthwise across the road; and no stone should be more than 5 inches broad on its face. "The irregularities of the upper part of the pavement should be broken off with the hammer, and all the interstices should be filled with stone chips, firmly wedged, or packed by hand with a light hammer; so that, when the pavement is finished, there may be a convexity of 4 inches in the breadth of 15 feet from the centre.

"The middle 18 feet of pavement should be coated with hard broken stones, of the form and size described under the head 'Road Materials,' to the depth of

6 inches. Four of these 6 inches to be first put on, and worked in by carriages and horses; care being taken to rake in the ruts until the surface becomes firm and consolidated, after which the remaining 2 inches are to be put on.

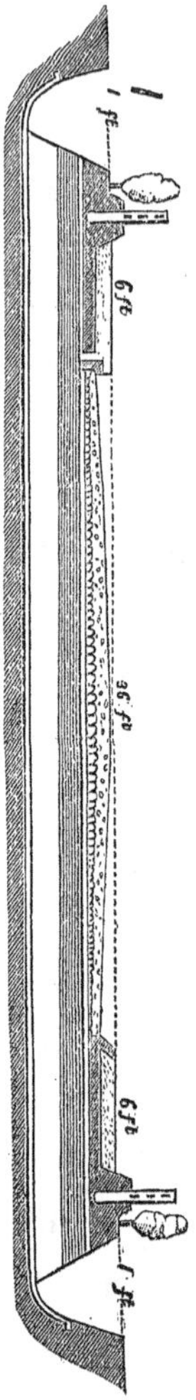

"The paved spaces on each side of the 18 middle feet should be coated with broken stones, or well-cleansed strong gravel, up to the footpath, or other boundary of the road, so as to make the whole convexity of the road 6 inches from the centre to the sides of it; and the whole of the materials should be covered with a binding of an inch and a half in depth of good gravel, free from clay or earth."

The footpaths should be made with a coating of strong gravel, or small broken stones, at least 6 inches deep. The annexed engraving exhibits a section of a road constructed according to the above rules.

REPAIRING ROADS.

Towards the latter end of the autumn of each year, a road should be put into a complete state of repair, to preserve it from being broken up during the following winter, between which time and the preceding spring, all repairs, by laying on of new materials, should be done. If thin coatings be laid on at a time, and when the ground is wet, they will soon be worked into the surface without being crushed into powder.

All ruts and hollows should be filled up as soon as they appear. The side channels and drains should be continually kept clean, and free from obstruction ; and all damage they may have sustained be made good as soon as discovered.

" A road should be scraped from time to time, so as never to have half an inch of mud on it ; the mud should not be scraped into, or allowed to remain in, the side channels, so as to stop the running of water in them.

" The hedges should be kept constantly clipped and cut as low as possible, without rendering them unfit for confining cattle ; and all projecting branches of the trees in the fences should be lopped."

In the minutes of evidence given before a Select Committee of the House of Commons on the subject of Steam Carriages, we find the following paragraph as part of the evidence given by Sir John Macneill :

"Well-made roads, formed of clean, hard, broken stone, placed on a solid foundation, are very little affected by changes of atmosphere ; weak roads, or those that are imperfectly formed of gravel, flint, or round pebbles, without a bottoming or foundation of stone pavement or concrete, are, on the contrary, much affected by changes of the weather. In the formation of such roads, and before they become bound or firm, a considerable portion of the subsoil mixes with the stone or gravel, in consequence of the necessity of putting the gravel on in thin layers: this mixture of earth or clay, in dry warm seasons, expands by the heat, and makes the road loose and open; the consequence is, that the stones are thrown out, and many of them are crushed and ground into dust, producing considerable wear and diminution of the materials: in wet weather, also, the clay or earth, mixed with the stones, absorbs moisture, becomes soft, and allows the stones to move and rub against each other when acted upon by the feet of horses or wheels of carriages. This attrition of the stones against each other wears them out surprisingly fast, and produces large quantities of mud, which tend to keep the road damp, and by that means increase the injury."

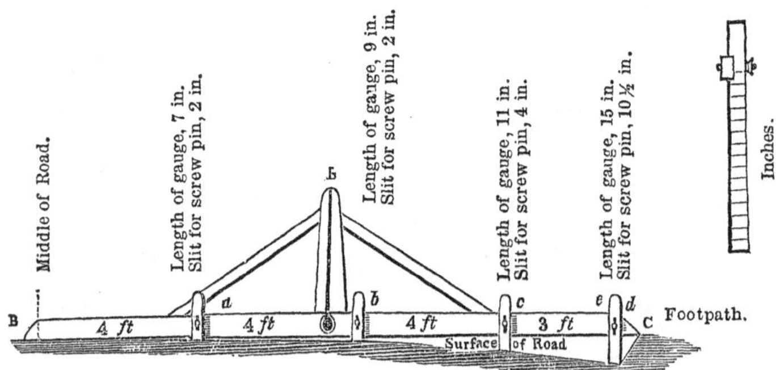

The above engraving represents the level employed by road-surveyors in laying out new works. On the horizontal bar A C are placed four sliding gauges, *a*, *b*, *c*, *d*, which move in dovetailed grooves cut in the horizontal bar, and when adjusted to their proper depth below the bottom edge of the level, can be firmly fixed in their position by a thumb-screw. A section of this portion of the instrument, taken through the line at *e*, is given on the right, drawn to a larger scale ; the remaining parts of the instrument require no explanation.

For laying out slopes, the clinometer, described at page 96, is the best instrument that can be used.

TABLE I.—*Showing the reduction upon each chain necessary to reduce hypothenusal to horizontal measure.*

Angle of ascent or descent.		Reduction in links.	Angle of ascent or descent.		Reduction in links.	Angle of ascent or descent.		Reduction in links.
°	′		°	′		°	′	
4	0	¼	23	48	8½	33	55	17
5	44	½	24	30	9	34	25	17½
7	2	¾	25	11	9½	34	55	18
8	7	1	25	51	10	35	25	18½
11	28	2	26	30	10½	35	54	19
12	50	2½	27	8	11	36	24	19½
14	5	3	27	45	11½	36	53	20
15	13	3½	28	22	12	37	21	20½
16	15	4	28	58	12½	37	49	21
17	15	4½	29	33	13	38	17	21½
18	12	5	30	8	13½	38	45	22
19	6	5½	30	41	14	39	12	22½
19	57	6	31	15	14½	39	39	23
20	47	6½	31	48	15	40	6	23½
21	34	7	32	20	15½	40	33	24
22	20	7½	32	52	16	40	58	24½
23	5	8	33	24	16½	41	25	25

TABLE II.—*Gradients or Inclined Planes.*

Ascent or Descent.		Rate of inclination.	Angle of inclination.	Ascent or Descent.		Rate of inclination.	Angle of inclination.
In 1 mile.	In 1 chain.			In 1 mile.	In 1 chain.		
feet.	ft. dec.		° ′ ″	feet.	ft. dec.		° ′ ″
1	0.013	1 in 5280.0	0 0 39	51	0.638	1 in 103.5	0 33 12
2	0.025	.. 2640.0	0 1 18	52	0.650	.. 101.5	0 33 51
3	0.038	.. 1760.0	0 1 57	53	0.663	.. 99.6	0 34 30
4	0.050	.. 1320.0	0 2 36	54	0.675	.. 97.8	0 35 10
5	0.063	.. 1056.0	0 3 16	55	0.688	.. 96.0	9 35 49
6	0.075	.. 880.0	0 3 55	56	0.700	.. 94.3	0 36 28
7	0.088	.. 754.3	0 4 34	57	0.713	.. 92.6	0 37 7
8	0.100	.. 660.0	0 5 13	58	0.725	.. 91.0	0 37 46
9	0.113	.. 586.7	0 5 52	59	0.738	.. 89.5	0 38 25
10	0.125	.. 528.0	0 6 31	60	0.750	.. 88.0	0 39 4
11	0.138	.. 480.0	0 7 10	61	0.763	.. 86.6	0 39 43
12	0.150	.. 440.0	0 7 49	62	0.775	.. 85.2	0 40 22
13	0.163	.. 406.1	0 8 28	63	0.788	.. 83.8	0 41 1
14	0.175	.. 377.1	0 9 7	64	0.800	.. 82.5	0 41 40
15	0.188	.. 352.0	0 9 46	65	0.813	.. 81.2	0 42 19
16	0.200	.. 330.0	0 10 25	66	0.825	.. 80.0	0 42 58
17	0.213	.. 310.6	0 11 4	67	0.838	.. 78.8	0 43 37
18	0.225	.. 293.3	0 11 43	68	0.850	.. 77.6	0 44 16
19	0.238	.. 277.9	0 12 22	69	0.863	.. 76.5	0 44 55
20	0.250	.. 264.0	0 13 1	70	0.875	.. 75.4	0 45 34
21	0.263	.. 251.4	0 13 40	75	0.938	.. 70.4	0 48 50
22	0.275	.. 240.0	0 14 20	80	1.000	.. 66.0	0 52 5
23	0.288	.. 229.6	0 14 59	85	1.063	.. 62.1	0 55 20
24	0.300	.. 220.0	0 15 38	90	1.126	.. 58.7	0 58 36
25	0.313	.. 211.2	0 16 17	95	1.188	.. 55.6	1 1 51
26	0.325	.. 203.1	0 16 56	100	1.250	.. 52.8	1 5 6
27	0.338	.. 195.6	0 17 35	110	1.375	.. 48.0	1 11 37
28	0.350	.. 188.6	0 18 14	120	1.500	.. 44.0	1 18 7
29	0.363	.. 182.1	0 18 53	130	1.625	.. 40.6	1 24 38
30	0.375	.. 176.0	0 19 32	140	1.750	.. 37.7	1 31 8
31	0.388	.. 170.3	0 20 11	150	1.875	.. 35.2	1 37 38
32	0.400	.. 165.0	0 20 50	160	2.000	.. 33.0	1 44 8
33	0.413	.. 160.0	0 21 29	170	2.125	.. 31.1	1 50 39
34	0.425	.. 155.3	0 22 8	180	2.250	.. 29.3	1 57 9
35	0.438	.. 150.9	0 22 47	190	2.375	.. 27.8	2 3 39
36	0.450	.. 146.7	0 23 26	200	2.500	.. 26.4	2 10 9
37	0.463	.. 142.7	0 24 5	220	2.750	.. 24.0	2 23 9
38	0.475	.. 138.9	0 24 44	240	3.000	.. 22.0	2 36 9
39	0.488	.. 135.4	0 25 23	260	3.250	.. 20.3	2 49 9
40	0.500	.. 132.0	0 26 3	280	3.500	.. 18.9	3 2 8
41	0.513	.. 128.8	0 26 42	300	3.750	.. 17.6	3 15 7
42	0.525	.. 125.7	0 27 21	320	4.000	.. 16.5	3 28 6
43	0.538	.. 122.8	0 28 0	340	4.250	.. 15.3	3 41 4
44	0.550	.. 120.0	0 28 39	360	4.500	.. 14.7	3 54 2
45	0.563	.. 117.3	0 29 18	380	4.750	.. 13.9	4 6 59
46	0.575	.. 114.8	0 29 57	400	5.000	.. 13.2	4 19 56
47	0.588	.. 112.3	0 30 36	425	5.313	.. 12.4	4 36 7
48	0.600	.. 110.0	0 31 15	450	5.625	.. 11.7	4 52 17
49	0.613	.. 107.8	0 31 54	475	5.938	.. 11.1	5 8 26
50	9.625	.. 105.6	0 32 33	500	6.250	.. 10.6	5 24 35

APPENDIX.

In the former edition of this work there was an error in the Rule which appears, in the present edition, at page 75, the word "tangent" having been written for "sine." A scientific friend, in noticing this mistake, has suggested a convenient method of arranging the figures of the calculation when the perpendicular and base of a right-angled triangle are *both* to be computed from the base-angle and hypothenuse. The directions are as follows :—

"1st. Write down the log. of the measured hypothenuse taken from the table of numbers.

"2nd. Over it place the log. sine of the measured angle from table of log. sines, &c., and draw a line *above*.

"3d. Under the log. of the hypothenuse write down the cosine of the angle, and then draw a line *under* it. Add the hypothenuse to the sine *upwards*, and it will give the length of the perpendicular sought in the table of numbers. Add the hypothenuse and cosine together *downwards*, and it will give the length of the base in the table of numbers.

"NOTE.—I reject 10 from the log. index in both sums, beeause radius 10 stood in the first term, as a divisor, in both proportions.

"EXAMPLE.

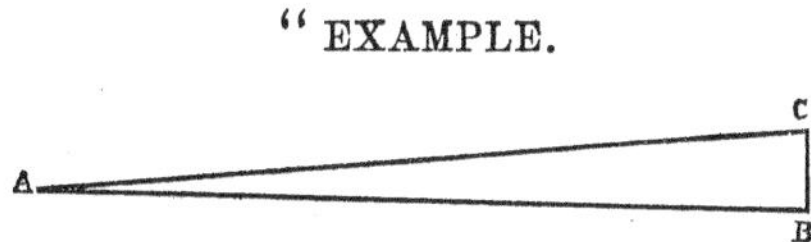

"From A to C, I found the angle to be 6° rising—and the length of the hypothenuse A C measured 1240 links.

"I state the calculation thus:

		2.112657 = log $129\frac{6}{10}$ the perp^r.
2nd, Sine of 6°	=	9.019235
1st, Hypothenuse 1240	=	3.093422
3rd, Cosine of 6°	=	9.997614
		3.091036 = log $1233\frac{2}{10}$ the base.

Answer:

Perpendicular . .	$129\frac{6}{10}$	links.
Base	$1233\frac{2}{10}$	do.
Hypothenuse . .	1240	do.

We may add here, in reference to the precept in italics, at page 75, that the constant number 1.8195439 is the logarithm of 66, the number of feet in a chain.

ON SETTING OUT THE WIDTHS OF GROUND REQUIRED

FOR THE WORKS OF A

RAILWAY OR CANAL,

ETC., ETC.,

DEPENDING UPON THE DEPTH OF CUTTING OR
HEIGHT OF EMBANKMENT,
AND THE TRANSVERSE SLOPE OF THE NATURAL SURFACE.

BY

FREDERICK WALTER SIMMS, F. G. S., M. I. C. E.

8

ON SETTING OUT THE WIDTHS OF GROUND REQUIRED

FOR THE WORKS OF A

RAILWAY OR CANAL,

ETC., ETC.

When the natural surface of the ground, both longitudinally and transversely, is upon the same level as that of the intended works, the process of setting and staking out the widths is very simple. Let us take, for example, the case of a railway, the base or bottom width of which, when prepared for the reception of the ballasting and permanent way, is to be 36 feet; the ratio of the inclination, or batter, of the slopes to the heights, both in the cuttings and the embankments, to be 2 to 1; beyond which, or at the outward edge, a slip of land 12 feet wide is to be taken on each side of the railway for the fences, &c. First, the centre line must be staked out and carefully levelled: it is customary to drive a stake, about 2 feet long and about 1½ inch square, into the ground at each chain's length, their tops to be upon the fair level of the natural surface, thus affording good stations for the levelling staves to be held upon; the relative level of each stake being then very accurately determined with respect to some given datum, they become so many zero points for reference in the subsequent operations. From each of the centre stakes a line must be set out on both sides, and at right angles to the centre line, or at right angles to a tangent to the centre line at that point, if the centre line be curved: upon these transverse lines the required widths of land must be set out. Now, if the ground at any of the centre stakes is upon the same level as the intended base of the railway, nothing more will be required than to measure on each transverse line, and in both directions from the centre stake, one half the required width, which, in our supposed case, is 18 feet for the half width of the railway, and 12 feet for the fences; in all 30 feet on each side of the centre. But when, as it mostly happens, the ground is not on the proposed level of the railway, the operation is not quite so simple; and if in addition thereto the ground slopes sidewise or at right angles to the general direction of the line, the business is still more complicated, and requires some skill and care to do the work correctly. The method of doing this it is now our business to explain.

The next most simple case to the above is when the cross section of the ground is horizontal, be the depth of cutting or height of embankment what it may.

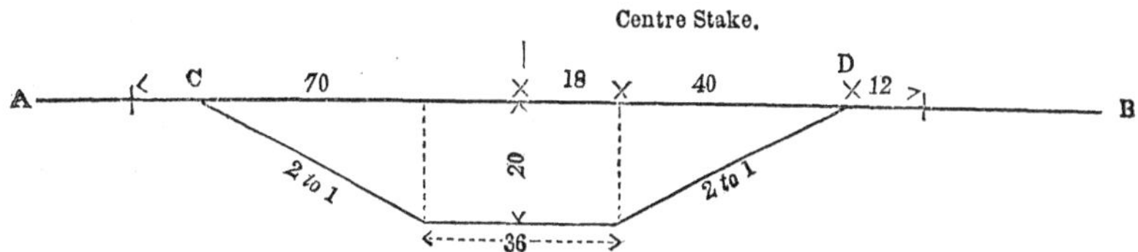

This is shown in the preceding diagram, which represents a cross section of a 20 feet cutting, with slopes of two horizontal to one perpendicular. The horizontal line A B at right angles to the centre line represents the natural surface of the ground. Under these circumstances it will readily be seen that the half width of the cutting, or the distance from the centre to the edge of the slopes C and D, equals the half width of the base (18) added to the batter of the sloping sides (40), and including the 12 feet for the fences, the total half width of land required for the purposes of such railway would be $18 + 40 + 12 = 70$ feet, and consequently the whole required width to be appropriated and fenced in for a 20 feet cutting or embankment, when the ground does not slope sidewise, would be 140 feet.

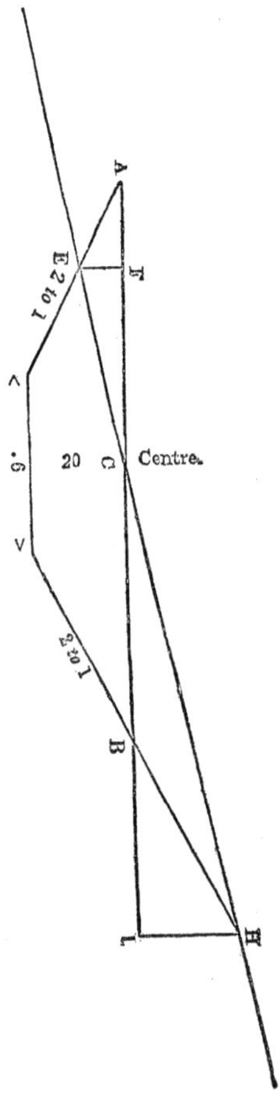

The next and more complicated, and also the most frequently occurring case, is, when the cross section of the natural surface is not horizontal, as shown in the annexed diagram, which also represents a cutting of 20 feet.

Let the line A B represent a horizontal line passing

through the centre line C of the railway, which, if it coincided with the surface of the ground, would give A C and C B (each half width) 70 feet, as in the former example, the depth of cutting and the slopes being assumed the same.

Let the line E H represent the natural surface of the ground upon this transverse section ; it will readily be perceived that the real half width C E (on the left of the diagram) is much shorter than the horizontal or computed half width A C, because the ground-line is depressed on that side of the centre ; likewise the half width C H on the other side of the centre is greater than the said horizontal or computed half width, because the ground is there elevated above the horizontal line A B passing through the centre. To determine *exactly* the distances C E and C H in actual operations in the field, would be attended with some difficulty, and consume much time ; but the following method, at the same time that it gives a sufficiently correct approximation in practice, is also a very expeditious one :

Let us suppose that the point E or distance C E be known, and that with a spirit-level we determine the difference of level between the points C and E, this difference is represented by the line E F, which suppose to be one foot ; now we have a small right-angled triangle A E F, of which E F is determined, being the difference of level (one foot), and the slope or ratio of A F to E F also given (2 to 1), therefore the side A F is known (2 feet), which, subtracted from the computed half width A C, leaves F C approximately equal to E C, the required half width, sufficiently exact for all practical purposes, where the cross section of the ground does not differ materially from a horizontal line.

We have been supposing that the point E is known, whereas that point is the object of our search; in practice, therefore, we proceed thus:—Take the computed half width, and if the ground is *depressed*, let a levelling staff be held somewhat *nearer* the point C than the said computed half width, for a first approximation to the point E; then determine the *difference of level* between this assumed point and the centre point C, *multiply this difference of level by the ratio of the slopes* (which doubles it when the slope is 2 to 1), and *subtract* the result from the computed half width, which gives a more correct approximation to the point E; now hold the staff at this *new point* and find the difference of level as before, again multiply by the ratio of the slopes, and deduct the result from the computed half width, which second result will in most cases be sufficiently near the real half width for a *depressed* line for all practical purposes.

Example.—Central height (or depth of cutting), 20 feet, slopes 2 to 1, base 36 feet; the computed half width was therefore 58 feet; the ground being depressed, we estimated that the point E might fall short of the computed half width 2 feet: we therefore directed a levelling staff to be held at 56 feet from the centre line (or stake) C, at which point another staff was held, and, by means of a spirit-level set up at a convenient distance, we found the difference of level between these points to be 0.87 foot, which, multiplied by the ratio of the slopes (2 to 1), gave 1.74 foot to be subtracted from the computed half width, 58 feet, leaving 56.26 feet for a first approximation to the half width C E (see last diagram). Now, upon removing the staff to this new point, the difference of level was again taken (or rather we should say that the staff was again read off, as the level had not been disturbed), and found to be 0.91 foot, which also multiplied by the ratio of the slopes (2 to 1), gave 1.82 foot to be substracted from 58 feet, leaving 56.18 for the second approximation, and which was adopted as the correct half width for the depressed side of the centre; indeed, in such a case as is above given, where the ground is so nearly horizontal, the first approximation (taken by a person after a little practice) may be assumed as the correct result, for in the above example it differed but .08 from the second determination, and if it had been taken a third time it could not have been more accurate as far as practice is concerned; this, however, is not the case where the inclination or slope of the ground is considerable, for then (if this method be followed) several approximations will be necessary to bring the result within admissible limits.

When the ground is *elevated* above the horizontal line, as shown on the right hand of the diagram, the mode of procedure will somewhat differ : thus, instead of holding the staff and finding the difference of level at a *less distance* than the computed half width, it must be held at a *greater distance* to obtain the point H by approximation ; the difference of level between that point and the centre point C being equal to H I, which multiplied by the ratio of the slopes, will give the distance B I to be added to the computed half width C B, to obtain the half width C H ; this may likewise be repeated to obtain a more correct result, as described for the other or depressed side of the centre C. It will also here be obvious to a person possessing but the smallest share of mathematical knowledge, that this result is not strictly correct, inasmuch as the line C H can never be equal to C I, but for practical purposes it is, as before observed, sufficiently correct. It may not be altogether unnecessary to observe, in this place, that the corrections B I, &c., as shown in the foregoing diagrams, are much exaggerated, being far greater in proportion to the computed half width C B, than ever occurs in ordinary practice, but this has been done to make our explanations more distinct than they otherwise would be.

The above particulars have been confined to the case of excavations ; we must now show in what the process differs when the ground is to be covered with an embankment.

By reversing page 120 we invert the diagram, which then represents an embankment. The rule for finding the half width for an embankment where the transverse section of the ground is horizontal, remains the same as for the cuttings under like circumstances, as may be seen by an inspection of the inverted figure of the first diagram; but upon inverting the second diagram, it will at once be seen that some variation in the process is required. Thus:

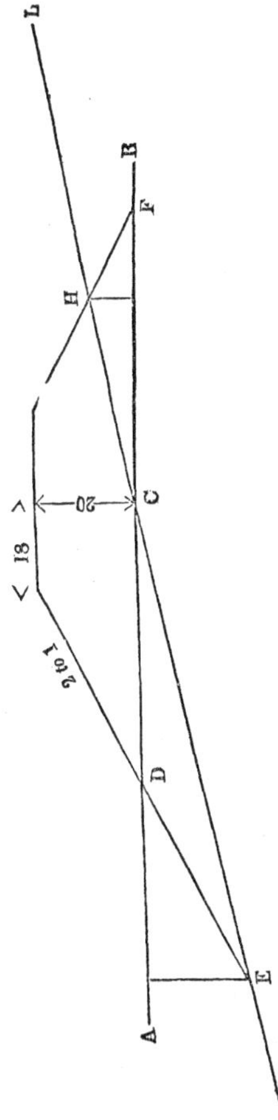

The horizontal line is represented by that marked A B; C D and C F are the computed half widths; C E the required half width on the depressed side, and C H the required half width on the elevated side, the line K L representing the natural surface of the ground. In the case of an excavation, we have shown that the *real* half width is greater on the *elevated* side than the computed half width, and less on the *depressed* side; but it will be seen by the above diagram that for an embankment the *real* half width is *less* on the *elevated* side, and *greater* on the *depressed* side, than the computed half width; therefore, in determining the approximate place of the point E on the depressed side for an embankment, the staff must be held *further* from the centre than the computed half width; and for the point H, or the elevated side, it must be held *nearer* to the centre than the computed half width; and finally, for computing the real half widths from the differences of level between the points E and the centre, and H and the centre; on the *depressed* side the difference of level multiplied by the ratio of the slopes is to be *added* to the computed half widths to obtain the point E, and to be *subtracted* from the computed half widths to obtain the point H.

The process above described may appear to the reader a very tedious one; it perhaps is so to read; but a little practice will convince him that it is a very expeditious method, for in most cases one setting up of the level will answer for several stations, and the multiplication by the ratio of the slopes upon such small numbers as mostly occur is easily performed, especially if it be an even number, as 2 to 1. The columns of the field-book may be arranged as in the following example for making the calculations in the field, or may be abridged to suit a more convenient-sized book for the pocket, at the pleasure of the surveyor; indeed, all that can be accomplished now of this kind is to give general rules, which can be altered and arranged to suit the convenience of the surveyor, as experience may point out a more suitable mode of proceeding. The example is taken from an extensive field operation by the writer, and shows the work both for a cutting and an embankment; the change from one to the other, or the tailing out of the cutting, as it is called, being included therein. The slope of the cutting is calculated at 1½ to 1, and that of the embankment at 2 to 1. The width of the railway was 36 feet, consequently half the said width was 18 feet.

EXAMPLE.

No. of Stake.	Depth of Cutting or Embankment.	Computed half width.	Section or Level Readings at right angles to Line.			Difference of Level. ±		Difference of Level, × ratio of Slope. ±		Required half width for edge of Cutting or foot of Embankment.	
			South.	Centre.	North.	South.	North.	South.	North.	South.	North.
EMBANKMENT.	Feet.	Feet.	Feet.	Feet.	Feet.	Feet.	Feet.	Feet.	Feet.	Feet.	Feet.
285	16.97	51.94	10.90	7.50	3.96	+ 3 40	− 3.54	+ 6.80	− 7.08	58.74	44.86
286	1.43	0.86	7.06	4.74	3.24	+ 2.32	{ + 0.07 − 1.50 }	+ 4.64	+ 0.11	25.50	20.97
287	2.77	23.54	8.00	5.80	4.26	+ 2.20	− 1.54	+ 4.40	− 3.08	27.94	20.46
288	3.06	24.12	8.82	6.42	5.12	+ 2.40	− 1.30	+ 4.80	− 2.60	28.92	21.52
289	2.01	22.02	7.02	5.13	3.74	+ 1.89	− 1.39	+ 3.78	− 2.78	25.80	19.24
290	1.22	20.44	6.00	4.10	2.76	+ 1.90	{ + 0.12 − 1.34 }	+ 3.80	+ 0.18	24.24	20.62
291	1.91	21.82	7.52	6.95	5.20	+ 0.57	− 1.75	+ 1.14	− 3.50	22.96	18.32
CUTTING.											
292	1.39	20.78	12.20	11.35	10.52	− 0.85	+ 0.83	− 1.27	+ 1.24	19.51	22.02
293	4.51	27.02	9.56	7.98	6.22	− 1.58	+ 1.76	− 2.37	+ 2.64	24.65	29.66
294	5.72	29.44	8.40	6.52	4.27	− 1.88	+ 2.25	− 2.82	+ 3.37	26.62	32.81
295	6.85	31.70	7.06	5.10	3.02	− 1.96	+ 2.08	− 2.94	+ 3.12	28.76	34.82
296	8.61	35.22	7.53	5.28	2.76	− 2.25	+ 2.52	− 3.37	+ 3.78	31.85	39.00

The first column contains the number of the central stakes, reckoned from the commencement of the work, which are convenient for reference.

The second column contains the depth of cutting or the height of embankment, as the case may be, at that point on the centre line.

The third column, the computed half width from the centre line to the edge of the cutting, or foot of embankment, upon the supposition that the ground is horizontal at right angles to the centre line; this half width, as before explained (p. 121), is found by multiplying the central height by the ratio of the slopes, and adding to the product half the width at the base of the railway.

The fourth, fifth, and sixth columns contain the readings from the levelling staves at the centre stake, and at the approximate points E and H (see last diagram).

The seventh and eighth columns contain the differences of level between the centre stake and the above approximate points. These numbers are simply the differences of the quantities in the three preceding columns (except at stakes 286 and 290, which we will presently explain), and the signs + and − denote whether they are positive or negative quantities, as respects the centre and the approximate points E and H.

The ninth and tenth columns contain the differences of level (contained in columns 7 and 8) multiplied by the ratio of the slopes, and must have the same signs

+ or − as the corresponding numbers in the preceding columns.

The last two columns contain the final half widths obtained by adding or substracting, according to the prefixed signs + or −, the numbers in the two preceding columns to the computed half width contained in column 3.

After the explanations already given, the reader can find no difficulty in tracing the steps of the example, except perhaps with the stakes 286 and 290, where the difference of level on the north side is represented by two numbers bracketed together, one having the sign +, and the other − : for the stake 286 the real difference of level on the north side the centre is a rise of 1.50, that is, the approximate point H is 1.50 foot above the centre stake : but it happens that the height of the embankment itself at that point is to be but 1.43 foot (column 2) ; therefore the approximate point H is above the intended top of the embankment, and consequently will not represent the foot of an embankment, but the edge of a cutting, and therefore the calculation for the half width on the north side must be treated as for a cutting whose depth is equal to the *height of the approximate point* H *above the intended top of the embankment ;* or, in other words, the *excess* of the difference of level between the centre stake and the approximate point H, above the intended height of the embankment, is the quantity to be entered in the column (7 or 8) "Difference of Level," and to be computed as for a cutting instead of embankment. In the case of stake 286 this excess is 0.07, to which is prefixed the sign plus ; this sum multiplied by the ratio of the slope being additive (for a cutting) on the elevated side of the centre, as before explained.

For the stake 290, the north side of the line (column 6) is 1.34 higher than the centre stake, and it, being embankment, would have the sign – prefixed (as shown by the lower number, column 8) : but the central height of the embankment at that point is but 1.22 (column 2) ; therefore, 1.34 – 1.22 = 0.12, which is the depth of cutting on the elevated side, and when multiplied by the ratio of the slopes is to be added to the computed half width to obtain the correct result. When the surface of the ground is much inclined at right angles to the centre line, the numbers to be operated upon become proportionally large.

As it is a case of frequent occurrence that one side will be a cutting when the other is an embankment, we wish it to be well understood, and therefore annex the accompanying diagram to illustrate it.

The line F G represents the natural surface of the ground, A B the horizontal line at the centre stake, C D the intended height of the embankment, K L the width or base of the railway, 36 feet, part of which is an embankment and part a cutting ; the point E, or foot of the embankment, will be determined in the usual way, as explained at page 126 ; but the point H, which is to be the edge of the cutting, must be found by subtracting D C (the height of embankment) from H I (the difference of level); the remainder, H M (which is *the excess of the difference of level between the centre stake and the approximate point* H *above the intended height of embankment*), multiplied into the ratio of the slope, must be added to the computed half width, or, in other words, treated as for a cutting, to obtain the said point H, as before stated. By reversing the diagram the corresponding case will become evident; namely, when the centre line is in cutting, and one side on embankment, while the other is in excavation; and the mode of proceeding will at once strike the reader after perusing what we have above written.

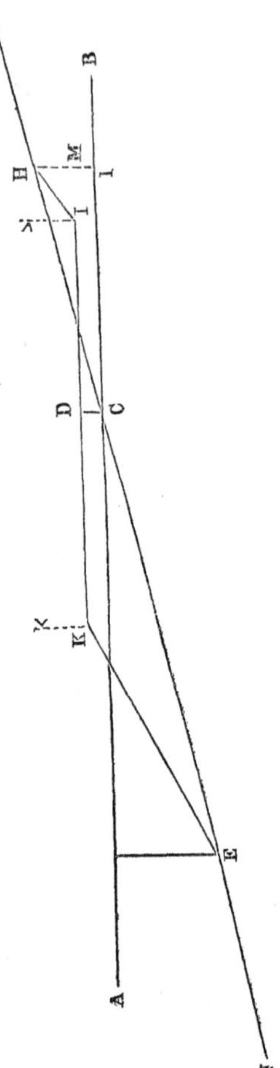

EXAMPLES OF THE MODES

OF SETTING OUT

RAILWAY CURVES.

By HENRY LAW.

EXAMPLES OF THE MODES

OF

SETTING OUT RAILWAY CURVES.

THERE are very few lines of railway so favorably situated as to be free from curves of greater or less extent, and occurring more or less frequently in their course ; and consequently a knowledge of the method of correctly and readily laying down these curves upon the ground becomes a very necessary and important qualification in those engaged in setting out lines of railway. It has not therefore been considered out of place to append to a work treating on one of the most important branches of railway surveying, a description of the various methods which may be employed for this purpose.

Previously, however, to proceeding to the more practical part of the subject, it may be desirable to make a few observations upon railway curves in general. The curve which has been almost universally employed in laying down lines of railway, is the arc of a circle, although it may be shown that, under certain circumstances, this curve is not *theoretically* that which should be employed, as affording the least danger from the centrifugal tendency of the carriages. It has been generally considered that the *true curve* was one which commenced with an infinite radius, decreasing in a regular manner in advancing on the curve, until the minimum radius of curvature required had been attained. This form of curve has, however, been deduced upon the assumption that the *whole* of the centrifugal tendency of the carriages is balanced by the superelevation of the outer rail, an assumption which is only correct on the supposition of the wheels of the carriages being cylindrical, or no play being allowed between the flanges and the rails—conditions which are never fulfilled in practice. For it may be shown, that with conical wheels, and a certain amount of play, a portion of the centrifugal tendency will always be counteracted by the self-acting adjustment produced by the lateral deviation of the carriage on the rails, however small the radius of curvature may be ; and that when the radius exceeds a certain limit, this adjustment is perfect, no superelevation of the outer rail being then required.

It is obvious, therefore, that in curves whose radii are *within* this limit, the true form for the curve is one whose radius of curvature at its commencement should equal this limit, and should decrease in advancing upon the curve, according to such a law, that (assuming the rise in the outer rail to form a regular inclined plane) the unbalanced centrifugal tendency should at every part of the curve be exactly counteracted by the amount of the superelevation of the rail at that part; until the top of the incline being reached, the radius of curvature should then remain constant, being such that the centrifugal tendency of the train should be exactly balanced by the combined effect of the lateral deviation of the carriages on the rails, and the superelevation of the outer rail. When, however, the radius of curvature exceeds this limit, it may be shown that the arc of a circle is preferable to any other form of curve.

Now, if we put d for the diameter of the wheels of the carriages, w for the width of the gauge of the line, and p for the lateral play allowed *on each side*, between the flanges of the wheels and the rails (all the dimensions being expressed in feet), and $\frac{1}{n}$ being the ratio of inclination of the tire of the wheel; then

$$\frac{n\,d\,w}{4\,p} = \mathrm{R} \quad . \quad . \quad . \quad . \quad . \quad . \quad . \qquad \text{I.}$$

will be the limit above referred to; that is, R will be the least radius of curvature which may be used without the necessity of raising the outer rail. And for any other smaller radius, putting v for the velocity of the train in miles per hour, and r for the radius of the curve in feet; then

$$\frac{.782\,v^2\,(n\,d\,w - 4\,p\,r)}{n\,d\,r} = \varepsilon \quad . \quad . \qquad \text{II.}$$

will be the superelevation of the outer rail in inches,

which will be required for that radius, in order that the whole of the centrifugal tendency of the train may be destroyed.

Although we have thus shown that for all curves having a smaller radius than R, it would not be correct, *theoretically*, to employ the arc of a circle, it is nevertheless very questionable whether it would be advisable, *in practice* (except under peculiar circumstances), to substitute the theoretical curve in its stead, inasmuch as the circular arc possesses the practical advantage of being laid down upon the ground with far greater facility ; and the only real objection which can be made to its use—namely, that of requiring a sudden and instantaneous superelevation of the outer rail at the point where the curve commences—may in a great measure be removed by commencing to raise the rail before arriving at this point, and making the rise form a gradual inclined plane, whose summit shall be attained at the commencement of the curve. By the adoption of this plan, although the centrifugal tendency (as without it) commences suddenly, the counteracting force produced by the superelevation of the outer rail, at the same instant attains its maximum, and the two forces therefore balance each other. Whereas, without it, the sudden commencement of the centrifugal force, being entirely unopposed, would at first tend to throw the carriages off the line, until, by the gradual elevation of the outer rail, it had been entirely destroyed.

Having thus far pursued the inquiry as to which form of curve it is most expedient to employ in practice in laying down a line of railway, and having shown that with hardly an exception the arc of a circle is practically the best, we shall confine ourselves to describing a few of the most generally applicable methods by which the circular arc may be traced on the ground.

The first method which we shall describe is that which has in practice been perhaps the most extensively used, although it possesses some objections, which we shall point out in the sequel. Let A B and

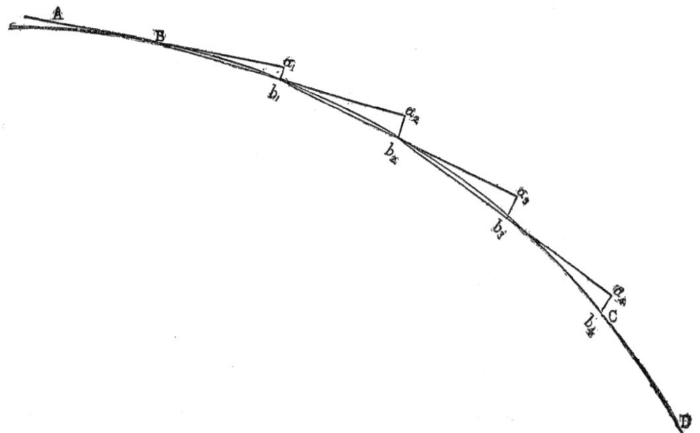

C D be the two straight portions of the line which it is desired to connect by a curve, B and C being the two points at which the curve falls into the straight lines; and let B b_1, b_1 b_2, b_2 b_3, &c., be the distance which it is desired that the points to be found in curve shall be apart: then measure upon the straight line A B produced, the distance B a_1, equal δ_1 in formula IV. below, and from the point a_1 set off, perpendicular to the same line, the distance a_1 b_1, equal to o_1 in formula III., which will give the first point required in the curve; then range a straight line through the points B, b_1, and upon this line lay off the distance b_1, a_2, equal to δ_2 in formula VI., and from the point a_2 set off, perpendicular to the line B a_2, the distance a_2 b_2, equal to o_2, in formula V., and the point b_2 will be the second point in the curve; then in a similar manner range another line through the points b_1 b_2, upon which measure the distance b_2 a_3, equal to the distance δ_2 or b_1 a_2, and from a_3 set off as before, perpendicular to the line b_1 a_3, the distance a_3 b_3, equal to o_2, which will determine the third point in the curve; and thus proceed until the whole extent of the curve has been set out.

In order to obtain the values of δ_1, δ_2, o_1, and o_2, let r equal the radius and d equal the distance B b_1, or b_1 b_2, &c., which it is desired that the points found in the curve shall be apart (both expressed in feet); then

$$\frac{d^2}{2\,r} = o_1 \quad . \quad . \quad . \quad . \quad . \quad . \quad . \quad . \quad \text{III.}$$

$$\sqrt{d^2 - {o_1}^2} = \delta_1 \quad . \quad . \quad . \quad . \quad . \quad . \quad \text{IV.}$$

$$\frac{d\,\delta_1}{r} = o_2 \quad . \quad . \quad . \quad . \quad . \quad . \quad . \quad . \quad \text{V.}$$

$$\frac{d\,(r - o_1)}{r} = \delta_2 \quad . \quad . \quad . \quad . \quad . \quad . \quad \text{VI.}$$

As an example of the application of this method, let the radius of the curve (r) be 15 chains or990 feet, and the distance B b_1 (d) one chain or 66 feet; then from formula III.,

$$\frac{66^2}{2 \times 990} = 2.2 \text{ feet} = o_1$$

will be the first offset at a_1; and

$$\sqrt{66^2 - 2.2^2} = 65.963 \text{ feet} = \delta_1$$

will be the distance B a_1, to be laid off upon the line A B produced to give the place for this offset. Again,

$$\frac{66 \times 65.963}{990} = 4.397 \text{ feet} = o_2$$

will be the offset at a_2, a_3, a_4, &c.; and

$$\frac{66 \times (990 - 2.2)}{990} = 65.85 \text{ feet} = \delta_2$$

will be the distance b_1 a_2, b_2 a_3, &c., to be measured from the points b_1, b_2, &c., in order to give the points a_2, a_3, a_4, &c., from which the offsets o_2 are to be taken.

To this method there are, as has been already stated, some practical objections, inasmuch as any error which may be committed, in setting out only one of the points in the curve, will occasion a corresponding error in every succeeding one; and a very trifling inaccuracy in calculating either the distance δ_2, or the length of the offset o_2, from its being frequently repeated, may ultimately cause a very considerable deviation from the true curve. Both these objections, however, may be in a great measure removed by the adoption of the following method of checking the position of about every fifth point; or, which would be better, first determining the position of these points, and then filling in the intermediate ones; and as we consider this modification does away almost entirely with the above-mentioned sources of error, we shall give an example of its application.

Let us suppose r and d, or the radius, and the distance the points B, b_1, b_2, &c., are apart, to be the

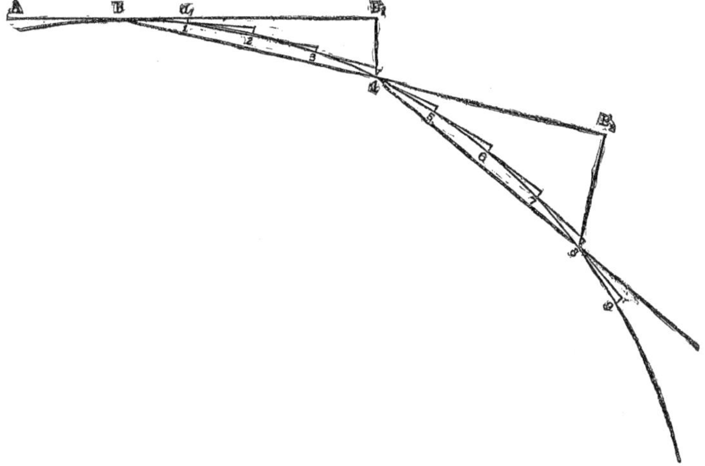

same as in the last example—viz., 990 feet and 66 feet respectively, and let it be determined to check the position of every *fourth* point: then the values of δ_1, δ_2, o_1, and o_2, will be the same as before; but previous to setting out the points 1, 2, 3, &c., we must calculate the distance B B_1 to be measured along the line A B produced, and the distance B_1 4 to be set off from the point B_1 to give the position of the fourth point in the curve, which may be done as follows: Let the distance B B_1 equal Δ_1, and B_1 4 equal O_1; and let D_1 be the length of the chord line connecting the two points B and 4 and β be the angle a B_1; then

$$\frac{o_1 \text{ rad}}{d} = \sin \beta,$$

and

$$\frac{2\,r \sin 4\,\beta}{\text{rad}} = D_1.$$

Then, by substituting D_1, O_1, and Δ_1, for d, o_1 and δ_1, in the formulæ III., IV., V., and VI., we shall obtain the values of O_1, Δ_1, O_2, and Δ_2 where Δ_2 is the distance 4 B_2 to be measured upon the chord line B_1 4 produced, and O_2 is the distance B_2 8 to be set off from B_2 in order to give the eighth point in the curve; for the values of r and d given above we shall obtain

Log o_1	= 0.342423	= log 2.2
Log rad	= 10.000000	
	10.342423	
Log d	= 1.819544	= log 66
Log sin β	= 8.522879	

and $\beta = 1° \; 54' \; 37'' \because 4\;\beta = 7° \; 38' \; 28''$; then

Log 2 r	= 3.296665	= log 1980
Log sin 4 β	= 9.123745	
	12.420410	
Log rad	= 10.000000	
Log D_1	= 2.420410	= log 263.27.

Then from formula III.

$$\frac{263.27^2}{2 \times 990} = 35 \text{ feet} = O_1;$$

from formula IV.

$$\sqrt{263.27^2 - 35^2} = 260.92 \text{ feet} = \Delta_1;$$

from formula V.

$$\frac{263.27 \times 260.92}{990} = 69.4 = O_2;$$

and from formula VI.

$$\frac{263.27 \times (990 - 35)}{990} = 253.96 = \Delta_2.$$

These being obtained, the position of every fourth point should be first determined by the dimensions Δ_1, O_1, Δ_2, and O_2; and then the intermediate points 1, 2, 3, 5, 6, &c., by δ_1, o_1, δ_2, and o_2, as first described.

The second method which we shall describe may be advantageously employed when the radius of curvature is large and the centre can be seen from every part of the curve.

Let the lines A B and C D as before represent the two straight portions of the line required to be connected by a curve having a radius of 80 chains or 1 mile. First set up a theodolite at B and another at C (the two terminations of the straight portions of the line) and from each point range a line at right angles to the lines A B and C D respectively, and at the intersection of these lines (E), which will be the centre of the curve, put up a signal sufficiently conspicuous to

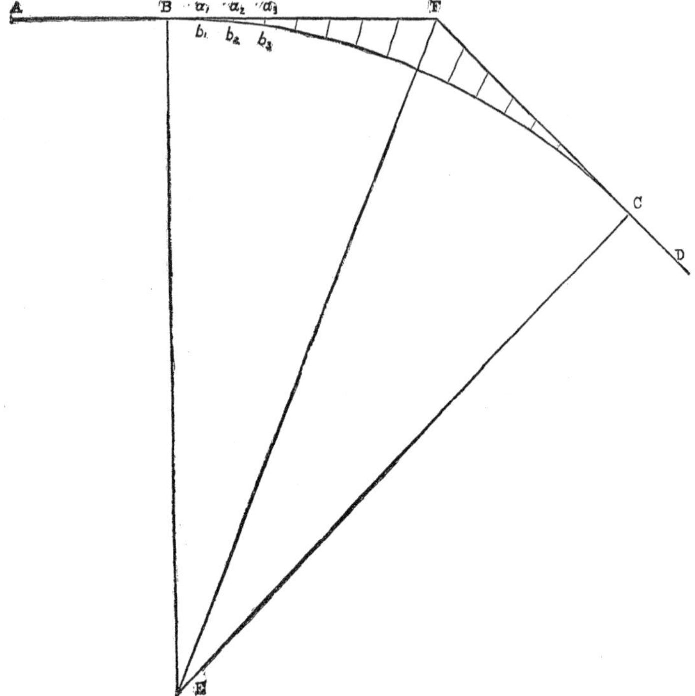

be seen from any point between B and C. Then produce the straight lines A B and C D until they intersect in the point F, and on these lines drive in stakes at equal distances, a_1, a_2, a_3, &c., commencing from the points B and C. If r equal the radius, and δ equal the distance between the points a_1, a_2, a_3, &c., both in feet, then

$$\sqrt{r^2+\delta^2}-r=o_1$$

will be the distance which must be set off from the

first point a_1, not perpendicular to the line B F, but in the direction a_1 E; in like manner

$$\sqrt{r^2 + \overline{2\,\delta}|^2} - r = o_2$$

will be the distance to be set off from the point a_2 in the direction a_2 E; and generally

$$\sqrt{r^2 + \overline{n\,\delta}|^2} - r = o_n$$

will be the distance to be set off at the nth points from B and C.

For example, let r be 5280 feet and δ equal 100 feet: then

$$\sqrt{5280^2 + 100^2} - 5280 = .94 \text{ feet} = o_1$$

will be the distance a_1 b_1, which must be set off from a_1 in the direction a_1 E to obtain the first point b_1 in the curve; and proceeding in a similar manner with the others, the following Table will exhibit the distances to be set off at the respective points a_1, a_2, a_3, &c.:

At a_1 or	100	feet from B,	the offset will be		.94
a_2	200	"		"	3.79
a_3	300	"		"	8.52
a_4	400	"		"	15.13
a_5	500	"		"	23.62
a_6	600	"		"	33.98
a_7	700	"		"	46.19
a_8	800	"		"	60·26
a_9	900	"		"	76.16
a_{10}	1000	"		"	93.86
a_{11}	1100	"		"	113.36

At a_{12}	or 1200 feet from B, the offset will be		134.65
a_{13}	1300 "	"	157.68
a_{14}	1400 "	"	182.45
a_{15}	1500 "	"	208.93

If the extent of the curve is such that the length of the offsets before reaching the point F, where the two tangent lines intersect, become inconveniently long, so as to occasion loss of time in setting them off, it will be advisable to make use of another tangent line as shown at G I; for determining the position of which

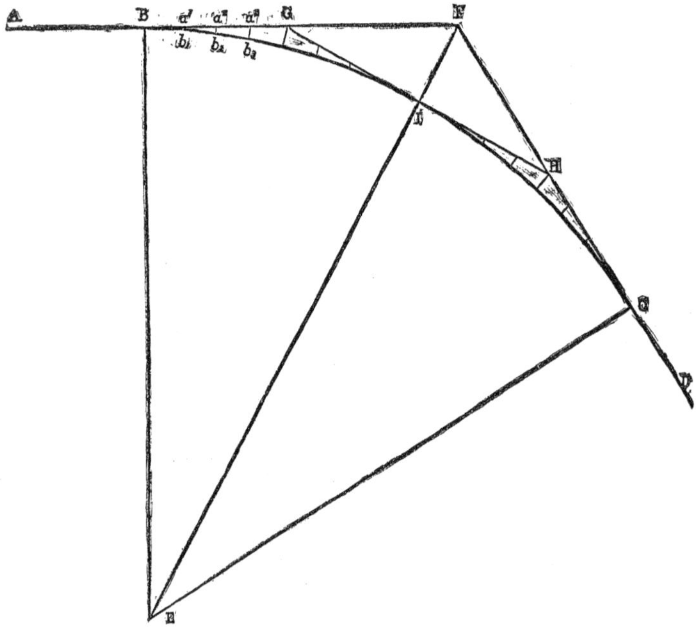

line the following method made be made use of. Let r, as before, be the radius, ε the number of degrees contained by the angle B E C, and n the number of tangent lines (as B G, G H, H I, I C) intended to be employed; then

$$\frac{r \sin \frac{\varepsilon}{n}}{\cos \frac{\varepsilon}{n}} = r \tan \frac{}{n}$$

will be equal to the length of any one of these tangent lines. As an example, let r be equal to 5280 feet, ε equal to 60°, and n equal to 4, so that the quotient of ε divided by n will be 15°: then the calculation for the length of each of the lines B G, G H, &c., will be as follows:—

$$\begin{aligned} \text{Log } r &= 3.722634 \\ \text{Log } \tan \frac{\varepsilon}{n} &= 9.428052 \\ & \overline{3.150686} = \log 1414 \end{aligned}$$

Hence the length of each of the lines B G, G H, &c., will be 1414.8 feet.

Now, having ascertained this length, nothing more remains than to set it off from B and C towards F, and then to range a line G I from the two points thus obtained, which will be the required tangent line: this line must then be bisected in the point H, which may readily be done by ranging a line from F to E, which having been done, proceed as already described to set off the equal distances a_1, a_2, a_3, &c., from B and H towards G, and from H and C towards I; and then by setting off the distances a_1, b_1, a_2, b_2, &c., contained in the table already given, from the several points a_1, a_2, &c., in directions radiating to the centre E, the course of the curve will be marked by the points b_1, b_2, b_3, &c., thus obtained.

One advantage possessed by the above method is, that, knowing exactly the direction in which to lay off the offsets (and that by the range of a comparatively distant object), the errors which have frequently arisen from their not having been set off perpendicularly, where the eye has been the only criterion, are entirely obviated; and this method is also entirely free from the objections made to the former method.

When the centre point E cannot be seen from every part of the curve, so as to allow the offsets being laid off radially, the more usual method may be adopted of laying off the offsets perpendicularly to the tangent B F, but in this case a cross staff should always be employed to insure accuracy, and the distances to be set off from the points a_1, a_2, a_3, &c., will be greater than those employed in the previous method, and must be calculated from the formula

$$r - \sqrt{r^2 - \delta^2} = 0_1$$

instead of that given at page 131.

The third method is most applicable where the radius of the curve is small as compared with its extent, and is deduced from the well-known theorem, that all angles contained in the same segment of a circle are equal to one another.* The method is as follows:— Place a theodolite at B and another at C, the two ter-

* Euclid, Book III., prop. 21.

minations of the straight portions of the line, setting the telescope of the instrument at B on C, and that at

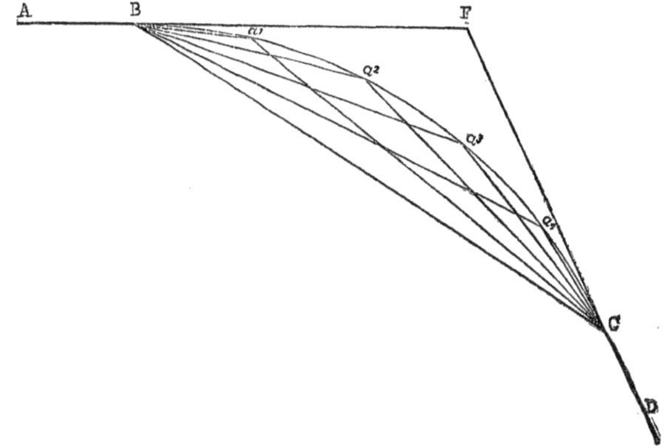

C on F, the point of intersection of the lines A B and C D produced; then if the former be moved through an arc of any number of degrees, towards F, and the latter the same number of degrees toward B, the point a_1, where the lines of collimation of the two telescopes intersect, will be a point in the curve; now let both theodolites be again moved the same number of degrees and in the same directions as before, and their axes produced, or lines of collimation, will again intersect at a_2, another point in the curve; and in fact, to whatever extent the theodolites are moved, so long as the arc described is equal in both, the point of their intersection will always be in the required curve. Or more generally suppose the two theodolites to be placed as first described, and then simultaneously to commence to revolve with the same uniform angular velocity, the point of intersection of their lines of collimation will describe the circular arc C, a_1, a_2, a_3, ... B ; and in equal intervals of time, equal portions of the arc will be described, which will be half as great as the arc, which would have been described in the same time, by the same angular velocity, at the centre of the circle (E) ; from which last-mentioned circumstance, we may readily calculate the magnitude of the angle through which the theodolites at B and C must be successively moved, in order that the points a_1, a_2, a_3, &c., at which their axes intersect, may be at the distance apart which it is desired that they should be. If r equals the radius of the curve, d the required distance, and β the angle a_1 B C ; then

$$\frac{d \text{ rad}}{2r} = \sin \beta \quad . \quad . \quad . \quad . \quad . \quad . \quad \text{VII.}$$

As an example of the application of this method, let r equal 20 chains, or 1320 feet, and let it be required to determine points in the curve at distances of about 100 feet ; now, from the above formula we shall obtain

$$\begin{aligned} \text{Log } d &= 2.000000 = \log 100 \\ \text{Log rad} &= 10.000000 \\ & \overline{12.000000} \\ \text{Log } 2r &= 3.421604 = \log 2640 \\ \text{Log } \sin\beta &= \overline{8.578396} \therefore \beta = 2^\circ\ 10'\ 15''. \end{aligned}$$

As it would be inconvenient, however, in practice, to lay off so frequently as would be required, an angle with odd minutes and seconds, we may instead of the above take an angle of 2 degrees, which will make the distance d equal 92.13 feet. Having thus determined the angle, and placed the theodolites as previously described—viz., that at B in the direction B C, and that at C in the direction C F—the former must be moved 2° towards F, and the latter 2° towards B, and a stake driven down at their point of intersection a_1; the former must then be moved 2° more towards F, and the latter 2° more towards B, and another stake put down at their point of intersection a_2, and so on until the theodolite at B is brought to the direction B F, and that at C to the direction C B, when the whole of the curve will have been staked out as required, the stakes being 92.13 feet apart. This method, the same as the last, is not liable to the objections that the first method was, and in addition possesses the very important practical advantage, that its accuracy is entirely independent of any undulation or change of level in the surface of the ground, an advantage which is not possessed by any of the other methods which we have described, the whole of which would require to have the distances and offsets corrected in proportion to the slope of the surface of the ground. In a hilly country —and it is in such districts that curves most frequently occur—this circumstance will render the last-described method far superior to either of those which precede it.

The next method which we shall give, is that described by Mr. Rankine, in a communication to the Institution of Civil Engineers, and depends on the theorem* that the angle subtended by any arc of a circle at the centre of a circle, is double the angle subtended by the same arc at any point in the circumference of the circle. The method of proceeding is as follows: first place a theodolite at B, the point where

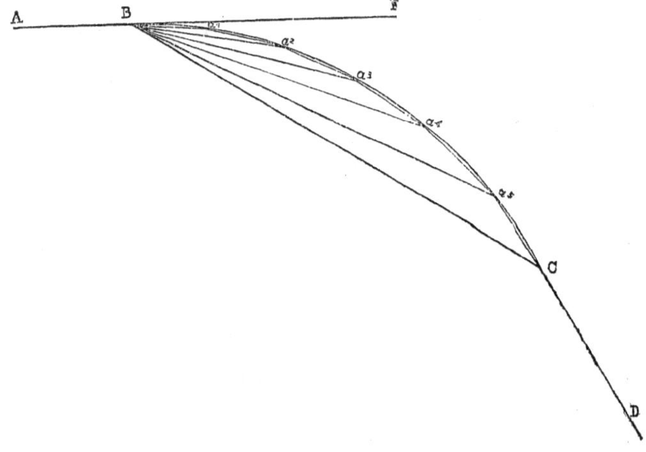

the curve commences; and then lay off from the line B F the angle B, calculated from formula VII. (supposing as before r to represent the radius of the curve and d the distance required between the points in the curve), and in the direction of the axis of the instrument set off the distance d, which will give the first point a_1 in the curve; in the same manner lay off from B F the angle 2 β, and from a_1 set off the same distance d, and the point where it cuts the axis of the instrument produced will be the second point a_2; and generally by laying off the angle $n\ \beta$, and setting off

* Euclid, Book III., prop. 20.

from the preceding point a_m the distance d, the point a_n will be given.

As an example of the application of this method, let r equal 19 chains, or 1254 feet, and d equal 100 feet; then from formula VII. we obtain

$$
\begin{array}{ll}
\text{Log } d = & 2.000000 = \log 100 \\
\text{Log rad} = & 10.000000 \\
\hline
 & 12.000000 \\
\text{Log } 2\,r = & 3.399328 = \log 2508 \\
\hline
\text{Log sin } \beta = & 8.600672 \because \beta = 2^\circ\ 17'\ 6'';
\end{array}
$$

then having placed the theodolite at the point B, lay off this angle 2° 17′ 6″ from the line B F, and upon the line B a_1 thus obtained set off 100 feet, which will give the first point in the curve a_1; then with an angle of 4° 34′ 12″ or 2 β set off another 100 feet from a_1, which will give the second point a_1, and thus proceed until the whole extent of the curve has been set out.

Having now pointed out several methods of procedure, in setting out the curved portions of a line of Railway, and having stated generally their relative advantages and disadvantages, we must leave it to the person using them to determine, from the circumstances attending any particular instance, which of these methods it would be preferable to employ. It may perhaps be necessary to add, that in passing from a curve of greater to one of less radius, and *vice versâ*, or in passing at once from a line curving in one direction to a line curving in the contrary direction, or a curve of contrary flexure, nothing more is requisite than to set off the tangent line to the curve at the point where the alteration occurs, and then to work from that line as from the line A B in any of the methods given above.

In conclusion, we would urge that too much care cannot be employed in the operations described above, much of the durability of the permanent way, freedom from jerks and uneasy motion, and also safety in travelling upon lines of Railway, depending upon the accuracy with which the rails are laid, the more especially on the curved portions of the line.

Scientific Books.

FRANCIS' (J. B.) Hydraulic Experiments. Lowell Hydraulic Experiments—being a Selection from Experiments on Hydraulic Motors, on the Flow of Water over Weirs, and in Open Canals of Uniform Rectangular Section, made at Lowell, Mass. By J. B. Francis, Civil Engineer. Second edition, revised and enlarged, including many New Experiments on Gauging Water in Open Canals, and on the Flow through Submerged Orifices and Diverging Tubes. With 23 copperplates, beautifully engraved, and about 100 new pages of text. 1 vol., 4to. Cloth. $15.

Most of the practical rules given in the books on hydraulics have been determined from experiments made in other countries, with insufficient apparatus, and on such a minute scale, that in applying them to the large operations arising in practice in this country, the engineer cannot but doubt their reliable applicability. The parties controlling the great water-power furnished by the Merrimack River at Lowell, Massachusetts, felt this so keenly, that they have deemed it necessary, at great expense, to determine anew some of the most important rules for gauging the flow of large streams of water, and for this purpose have caused to be made, with great care, several series of experiments on a large scale, a selection from which are minutely detailed in this volume.

The work is divided into two parts—Part I., on hydraulic motors, includes ninety-two experiments on an improved Fourneyron Turbine Water-Wheel, of about two hundred horse-power, with rules and tables for the construction of similar motors:—Thirteen experiments on a model of a centre-vent water-wheel of the most simple design, and thirty-nine experiments on a centre vent water-wheel of about two hundred and thirty horse-power.

Part II. includes seventy-four experiments made for the purpose of determining the form of the formula for computing the flow of water over weirs; nine experiments on the effect of back-water on the flow over weirs; eighty-eight experiments made for the purpose of determining the formula for computing the flow over weirs of regular or standard forms, with several tables of comparisons of the new formula with the results obtained by former experimenters; five experiments on the flow over a dam in which the crest was of the same form as that built by the Essex Company across the Merrimack River at Lawrence, Massachusetts; twenty-one experiments on the effect of observing the depths of water on a weir at different distances from the weir; an extensive series of experiments made for the purpose of determining rules for gauging streams of water in open canals, with tables for facilitating the same; and one hundred and one experiments on the discharge of water through submerged orifices and diverging tubes, the whole being fully illustrated by twenty-three double plates engraved on copper.

In 1855 the proprietors of the Locks and Canals on Merrimack River, at whose expense most of the experiments were made, being willing that the public should share the benefits of the scientific operations promoted by them, consented to the publication of the first edition of this work, which contained a selection of the most important hydraulic experiments made at Lowell up to that time. In this second edition the principal hydraulic experiments made there, subsequent to 1855, have been added, including the important series above mentioned, for determining rules for the gauging the flow of water in open canals, and the interesting series on the flow through a submerged Venturi's tube, in which a larger flow was obtained than any we find recorded.

FRANCIS (J. B.) On the Strength of Cast-Iron Pillars, with Tables for the use of Engineers, Architects, and Builders. By James B. Francis, Civil Engineer. 1 vol., 8vo. Cloth. $2.

HOLLEY'S RAILWAY PRACTICE. American and European Railway Practice, in the Economical Generation of Steam, including the materials and construction of Coal-burning Boilers, Combustion, the Variable Blast, Vaporization, Circulation, Superheating, Supplying and Heating Feed-water, &c., and the adaptation of Wood and Coke-burning Engines to Coal-burning; and in Permanent Way, including Road-bed, Sleepers, Rails, Joint Fastenings, Street Railways, &c., &c. By ALEXANDER L. HOLLEY, B. P. With 77 lithographed plates. 1 vol., folio. Cloth. $12.

"This is an elaborate treatise by one of our ablest civil engineers, on the construction and use of locomotives, with a few chapters on the building of Railroads. * * * All these subjects are treated by the author, who is a first-class railroad engineer, in both an intelligent and intelligible manner. The facts and ideas are well arranged, and presented in a clear and simple style, accompanied by beautiful engravings, and we presume the work will be regarded as indispensable by all who are interested in a knowledge of the construction of railroads and rolling stock, or the working of locomotives."—*Scientific American.*

HENRICI (OLAUS). Skeleton Structures, especially in their Application to the Building of Steel and Iron Bridges. By OLAUS HENRICI. With folding plates and diagrams. 1 vol., 8vo. Cloth. $3.

WHILDEN (J. K.) On the Strength of Materials used in Engineering Construction. By J. K. WHILDEN. 1 vol., 12mo. Cloth. $2.

"We find in this work tables of the tensile strength of timber, metals, stones, wire, rope, hempen cable, strength of thin cylinders of cast-iron; modulus of elasticity, strength of thick cylinders, as cannon, &c., effects of reheating, &c., resistance of timber, metals, and stone to crushing; experiments on brick-work; strength of pillars; collapse of tube; experiments on punching and shearing; the transverse strength of materials; beams of uniform strength; table of coefficients of timber, stone, and iron; relative strength of weight in cast-iron, transverse strength of alloys; experiments on wrought and cast-iron beams: lattice girders, trussed cast-iron girders; deflection of beams; torsional strength and torsional elasticity."—*American Artisan.*

CAMPIN (F.) On the Construction of Iron Roofs. A Theoretical and Practical Treatise. By FRANCIS CAMPIN. With wood-cuts and plates of Roofs lately executed. Large 8vo. Cloth. $3.

BROOKLYN WATER-WORKS AND SEWERS. Containing a Descriptive Account of the Construction of the Works, and also Reports on the Brooklyn, Hartford, Belleville, and Cambridge Pumping Engines. Prepared and printed by order of the Board of Water Commissioners. With illustrations. 1 vol., folio. Cloth. $15.

ROEBLING (J. A.) Long and Short Span Railway Bridges. By JOHN A. ROEBLING, C. E. Illustrated with large copperplate engravings of plans and views. Imperial folio, cloth. $25.

CLARKE (T. C.) Description of the Iron Railway Bridge across the Mississippi River at Quincy, Illinois. By THOMAS CURTIS CLARKE, Chief Engineer. Illustrated with numerous lithographed plans. 1 vol., 4to. Cloth. $7.50.

WILLIAMSON (R. S.) On the Use of the Barometer on Surveys and Reconnaissances. Part I. Meteorology in its Connection with Hypsometry. Part II. Barometric Hypsometry. By R. S. WILLIAMSON, Bvt. Lieut.-Col. U. S. A., Major Corps of Engineers. With Illustrative Tables and Engravings. Paper No. 15, Professional Papers, Corps of Engineers. 1 vol., 4to. Cloth. $15.

"SAN FRANCISCO, CAL., *Feb.* 27, 1867.

"Gen. A. A. HUMPHREYS, Chief of Engineers, U. S. Army:

"GENERAL—I have the honor to submit to you, in the following pages, the results of my investigations in meteorology and hypsometry, made with the view of ascertaining how far the barometer can be used as a reliable instrument for determining altitudes on extended lines of survey and reconnaissances. These investigations have occupied the leisure permitted me from my professional duties during the last ten years, and I hope the results will be deemed of sufficient value to have a place assigned them among the printed professional papers of the United States Corps of Engineers. Very respectfully, your obedient servant,

"R. S. WILLIAMSON,

"Bvt. Lt.-Col. U. S. A., Major Corps of U. S. Engineers."

TUNNER (P.) A Treatise on Roll-Turning for the Manufacture of Iron. By PETER TUNNER. Translated and adapted. By JOHN B. PEARSE, of the Pennsylvania Steel Works. With numerous engravings and wood-cuts. 1 vol., 8vo., with 1 vol. folio of plates. Cloth. $10.

SHAFFNER (T. P.) Telegraph Manual. A Complete History and Description of the Semaphoric, Electric, and Magnetic Telegraphs of Europe, Asia, and Africa, with 625 illustrations. By TAL. P. SHAFFNER, of Kentucky. New edition. 1 vol., 8vo. Cloth. 850 pp. $6.50.

MINIFIE (WM.) Mechanical Drawing. A Text-Book of Geometrical Drawing for the use of Mechanics and Schools, in which the Definitions and Rules of Geometry are familiarly explained; the Practical Problems are arranged, from the most simple to the more complex, and in their description technicalities are avoided as much as possible. With illustrations for Drawing Plans, Sections, and Elevations of Buildings and Machinery; an Introduction to Isometrical Drawing, and an Essay on Linear Perspective and Shadows. Illustrated with over 200 diagrams engraved on steel. By WM. MINIFIE, Architect. Seventh edition. With an Appendix on the Theory and Application of Colors. 1 vol., 8vo. Cloth. $4.

"It is the best work on Drawing that we have ever seen, and is especially a text-book of Geometrical Drawing for the use of Mechanics and Schools. No young Mechanic, such as a Machinist, Engineer, Cabinet-Maker, Millwright, or Carpenter should be without it."—*Scientific American.*

"One of the most comprehensive works of the kind ever published, and cannot but possess great value to builders. The style is at once elegant and substantial."—*Pennsylvania Inquirer.*

"Whatever is said is rendered perfectly intelligible by remarkably well-executed diagrams on steel, leaving nothing for mere vague supposition; and the addition of an introduction to isometrical drawing, linear perspective, and the projection of shadows, winding up with a useful index to technical terms."—*Glasgow Mechanics' Journal.*

☞ The British Government has authorized the use of this book in their schools of art at Somerset House, London, and throughout the kingdom.

MINIFIE (WM.) Geometrical Drawing. Abridged from the octavo edition, for the use of Schools. Illustrated with 48 steel plates. Fifth edition, 1 vol., 12mo. Half roan. $1.50.

"It is well adapted as a text-book of drawing to be used in our High Schools and Academies where this useful branch of the fine arts has been hitherto too much neglected."—*Boston Journal.*

PEIRCE'S SYSTEM OF ANALYTIC MECHANICS. Physical and Celestial Mechanics, by BENJAMIN PEIRCE, Perkins Professor of Astronomy and Mathematics in Harvard University, and Consulting Astronomer of the American Ephemeris and Nautical Almanac. Developed in four systems of Analytic Mechanics, Celestial Mechanics, Potential Physics, and Analytic Morphology. 1 vol., 4to. Cloth. $10.

GILLMORE. Practical Treatise on Limes, Hydraulic Cements, and Mortars. Papers on Practical Engineering, U. S. Engineer Department, No. 9, containing Reports of numerous experiments conducted in New York City, during the years 1858 to 1861, inclusive. By Q. A. GILLMORE, Brig.-General U. S. Volunteers, and Major U. S. Corps of Engineers. With numerous illustrations. One volume, octavo. Cloth. $4.

ROGERS (H. D.) Geology of Pennsylvania. A complete Scientific Treatise on the Coal Formations. By HENRY D. ROGERS, Geologist. 3 vols., 4to., plates and maps. Boards. $30.00.

BURGH (N. P.) Modern Marine Engineering, applied to Paddle and Screw Propulsion. Consisting of 36 colored plates, 259 Practical Woodcut Illustrations, and 403 pages of Descriptive Matter, the whole being an exposition of the present practice of the following firms: Messrs. J. Penn & Sons; Messrs. Maudslay, Sons, & Field; Messrs. James Watt & Co.; Messrs. J. & G. Rennie; Messrs. R. Napier & Sons; Messrs. J. & W. Dudgeon; Messrs. Ravenhill & Hodgson; Messrs. Humphreys & Tenant; Mr. J. T. Spencer, and Messrs. Forrester & Co. By N. P. BURGH, Engineer. In one thick vol., 4to. Cloth. $30.00. Half morocco. $35.00.

KING. Lessons and Practical Notes on Steam, the Steam-Engine, Propellers, &c., &c., for Young Marine Engineers, Students, and others. By the late W. R. KING, U. S. N. Revised by Chief-Engineer J. W. KING, U. S. Navy. Ninth edition, enlarged. 8vo. Cloth. $2.

WARD. Steam for the Million. A Popular Treatise on Steam and its Application to the Useful Arts, especially to Navigation. By J. H. WARD, Commander U. S. Navy. New and revised edition. 1 vol., 8vo. Cloth. $1.

WALKER. Screw Propulsion. Notes on Screw Propulsion, its Rise and History. By Capt. W. H. WALKER, U. S. Navy. 1 vol., 8vo. Cloth. 75 cents.

THE STEAM-ENGINE INDICATOR, and the Improved Manometer Steam and Vacuum Gauges: Their Utility and Application. By PAUL STILLMAN. New edition. 1 vol., 12mo. Flexible cloth. $1.

ISHERWOOD. Engineering Precedents for Steam Machinery. Arranged in the most practical and useful manner for Engineers. By B. F. ISHERWOOD, Civil Engineer U. S. Navy. With illustrations. Two volumes in one. 8vo. Cloth. $2.50.

POOK'S METHOD OF COMPARING THE LINES AND DRAUGHTING VESSELS PROPELLED BY SAIL OR STEAM, including a Chapter on Laying off on the Mould-Loft Floor. By SAMUEL M. POOK, Naval Constructor. 1 vol., 8vo. With illustrations. Cloth. $5.

SWEET (S. H.) Special Report on Coal; showing its Distribution, Classification and Cost delivered over different routes to various points in the State of New York, and the principal cities on the Atlantic Coast. By S. H. SWEET. With maps. 1 vol., 8vo. Cloth. $3.

ALEXANDER (J. H.) Universal Dictionary of Weights and Measures, Ancient and Modern, reduced to the standards of the United States of America. By J. H. ALEXANDER. New edition. 1 vol., 8vo. Cloth. $3.50.

"As a standard work of reference this book should be in every library; it is one which we have long wanted, and it will save us much trouble and research."—*Scientific American.*

CRAIG (B. F.) Weights and Measures. An Account of the Decimal System, with Tables of Conversion for Commercial and Scientific Uses. By B. F. CRAIG, M. D. 1 vol., square 32mo. Limp cloth. 50 cents.

"The most lucid, accurate, and useful of all the hand-books on this subject that we have yet seen. It gives forty-seven tables of comparison between the English and French denominations of length, area, capacity, weight, and the centigrade and Fahrenheit thermometers, with clear instructions how to use them; and to this practical portion, which helps to make the transition as easy as possible, is prefixed a scientific explanation of the errors in the metric system, and how they may be corrected in the laboratory."—*Nation.*

BAUERMAN. Treatise on the Metallurgy of Iron, containing outlines of the History of Iron manufacture, methods of Assay, and analysis of Iron Ores, processes of manufacture of iron and Steel, etc., etc. By H. BAUERMAN. First American edition. Revised and enlarged, with an appendix on the Martin Process for making Steel, from the report of Abram S. Hewitt. Illustrated with numerous wood engravings. 12mo. Cloth. $2.50.

"This is an important addition to the stock of technical works published in this country. It embodies the latest facts, discoveries, and processes connected with the manufacture of iron and steel, and should be in the hands of every person interested in the subject, as well as in all technical and scientific libraries."—*Scientific American.*

HARRISON. Mechanic's Tool Book, with practical rules and suggestions, for the use of Machinists, Iron Workers, and others. By W. B. HARRISON, associate editor of the "American Artisan." Illustrated with 44 engravings. 12mo. Cloth. $2.50.

"This work is specially adapted to meet the wants of Machinists and workers in iron generally. It is made up of the work-day experience of an intelligent and ingenious mechanic, who had the faculty of adapting tools to various purposes. The practicability of his plans and suggestions are made apparent even to the unpractised eye by a series of well-executed wood engravings."—*Philadelphia Inquirer.*

PLYMPTON. The Blow-Pipe: A System of Instruction in its practical use, being a graduated course of Analysis for the use of students, and all those engaged in the Examination of Metallic Combinations. Second edition, with an appendix and a copious index. By GEORGE W. PLYMPTON, of the Polytechnic Institute, Brooklyn. 12mo. Cloth. $2.

"This manual probably has no superior in the English language as a text-book for beginners, or as a guide to the student working without a teacher. To the latter many illustrations of the utensils and apparatus required in using the blow-pipe, as well as the fully illustrated description of the blow-pipe flame, will be especially serviceable."—*New York Teacher.*

NUGENT. Treatise on Optics: or, Light and Sight, theoretically and practically treated; with the application to Fine Art and Industrial Pursuits. By E. NUGENT. With one hundred and three illustrations. 12mo. Cloth. $2.

"This book is of a practical rather than a theoretical kind, and is designed to afford accurate and complete information to all interested in applications of the science."—*Round Table.*

SILVERSMITH (Julius). A Practical Hand-Book for Miners, Metallurgists, and Assayers, comprising the most recent improvements in the disintegration, amalgamation, smelting, and parting of the Precious Ores, with a Comprehensive Digest of the Mining Laws. Greatly augmented, revised, and corrected. By JULIUS SILVERSMITH. Fourth edition. Profusely illustrated. 1 vol., 12mo. Cloth. $3.

CLOUGH. The Contractors' Manual and Builders' Price-Book. By A. B. CLOUGH, Architect. 1 vol., 18mo. Cloth. 75 cents.

BRÜNNOW. Spherical Astronomy. By F. BRÜNNOW, Ph. Dr. Translated by the Author from the Second German edition. 1 vol., 8vo. Cloth. $6.50.

CHAUVENET (Prof. Wm.) New method of Correcting Lunar Distances, and Improved Method of Finding the Error and Rate of a Chronometer, by equal altitudes. By WM. CHAUVENET, LL.D. 1 vol., 8vo. Cloth. $2.

A SYNOPSIS OF BRITISH GAS LIGHTING, comprising the essence of the "London Journal of Gas Lighting" from 1849 to 1868. Arranged and executed by JAMES R. SMEDBERG, C. E. of the San Francisco Gas Works. Issued only to subscribers. 4to. Cloth. $15.00 *In press.*

GAS WORKS OF LONDON. By ZERAH COLBURN. 12mo. Boards. 60 cents.

HEWSON. Principles and Practice of Embanking Lands from River Floods, as applied to the Levees of the Mississippi. By WILLIAM HEWSON, Civil Engineer. 1 vol., 8vo. Cloth. $2.

"This is a valuable treatise on the principles and practice of embanking lands from river floods, as applied to Levees of the Mississippi, by a highly intelligent and experienced engineer. The author says it is a first attempt to reduce to order and to rule the design, execution, and measurement of the Levees of the Mississippi. It is a most useful and needed contribution to scientific literature."—*Philadelphia Evening Journal.*

WEISBACH (Julius). Principles of the Mechanics of Machinery and Engineering. By DR. JULIUS WEISBACH, of Freiburg. Translated from the last German edition. Vol. I. 8vo, cloth. $10.

HUNT (R. M.) Designs for the Gateways of the Southern Entrances to the Central Park. By RICHARD M. HUNT. With a description of the designs. I vol., 4to. Illustrated. Cloth. $5.

PEET. Manual of Inorganic Chemistry for Students. By the late DUDLEY PEET, M. D. Revised and enlarged by ISAAC LEWIS PEET, A. M. 18mo. Cloth. 75 cents.

WHITNEY (J. P.) Colorado, in the United States of America. Schedule of Ores contributed by sundry persons to the Paris Universal Exposition of 1867, with some Information about the Region and its Resources. By J. P. WHITNEY, of Boston, Commissioner from the Territory. Pamphlet. 8vo., with maps. London, 1867. 25 cents.

WHITNEY (J. P.) Silver Mining Regions of Colorado, with some account of the different processes now being introduced for working the Gold Ores of that Territory. By J. P. WHITNEY. 12mo. Paper. 25 cents.

"This is a most valuable little book, containing a vast amount of practical information about that region. It will be found useful to men of a scientific turn of mind, should they never contemplate a journey to the region of silver and gold."—*Fall River News.*

SILVER DISTRICTS OF NEVADA. 8vo., with map. Paper. 35 cents.

McCORMICK (R. C.) Arizona: Its Resources and Prospects. By Hon. R. C. McCORMICK. With map. 8vo. Paper. 25 cents.

PETERS. Notes on the Origin, Nature, Prevention, and Treatment of Asiatic Cholera. By JOHN C. PETERS, M. D. Second edition. With an appendix and map. 12mo. Cloth. $1.50.

SEYMOUR. Western Incidents connected with the Union Pacific Railroad. By SILAS SEYMOUR. 12mo. Cloth. $1.

EULOGIES IN MEMORY OF MAJ.-GEN. JAMES S. WADSWORTH AND COL. PETER A. PORTER, before the "Century Association." Tinted paper. 8vo. Paper. $1.

PALMER. Antarctic Mariners' Song. By JAMES CROXALL PALMER, U. S. N. Illustrated. Cloth, gilt, bevelled boards. $3.

"The poem is founded upon and narrates the episodes of the exploring expedition of a small sailing vessel, the 'Flying Fish,' in company with the 'Peacock,' in the South Seas, in 1838-42. The 'Flying Fish' was too small to be safe or comfortable in that Antarctic region, although we find in the poem but little of complaint or murmuring at the hardships the sailors were compelled to endure."—*Athenæum.*

FRENCH'S ETHICS. Practical Ethics. By Rev. J. W. FRENCH, D. D., Professor of Ethics, U. S. Military Academy. Prepared for the Use of Students in the Military Academy. I vol. 8vo. Cloth. $4.50.

AUCHINCLOSS. Application of the Slide Valve and Link Motion to Stationary, Portable, Locomotive, and Marine Engines, with new and simple methods for proportioning the parts. By WILLIAM S. AUCHINCLOSS, Civil and Mechanical Engineer. Designed as a hand-book for Mechanical Engineers, Master Mechanics, Draughtsmen, and Students of Steam Engineering. All dimensions of the valve are found with the greatest ease by means of a PRINTED SCALE, and proportions of the link determined *without* the assistance of a model. Illustrated by 37 woodcuts and 21 lithographic plates, together with a copperplate engraving of the Travel Scale. 1 vol. 8vo. Cloth. $3.

HUMBER'S STRAINS IN GIRDERS. A Handy Book for the Calculation of Strains in Girders and Similar Structures, and their Strength, consisting of Formulæ and Corresponding Diagrams, with numerous details for practical application. By WILLIAM HUMBER. 1 vol. 18mo. Fully illustrated. Cloth. $2.50.

GLYNN ON THE POWER OF WATER, as applied to drive Flour Mills, and to give motion to Turbines and other Hydrostatic Engines. By JOSEPH GLYNN, F. R. S. Third edition, revised and enlarged, with numerous illustrations. 12mo. Cloth. $1.25.

HOW TO BECOME A SUCCESSFUL ENGINEER; being Hints to Youths intending to adopt the Profession. By BERNARD STUART. Fourth edition. 18mo. Cloth. 75 cents.

"Parents and guardians, with youths under their charge destined for the profession, as well as youths themselves who intend to adopt it, will do well to study and obey the plain curriculum in this little book. Its doctrine will, we hesitate not to say, if practised, tend to fill the ranks of the profession with men conscious of the heavy responsibilities placed in their charge."—*Practical Mechanic's Journal.*

TREATISE ON ORE DEPOSITS. By BERNHARD VON COTTA, Professor of Geology in the Royal School of Mines, Freidberg, Saxony. Translated from the second German edition, by FREDERICK PRIME, Jr., Mining Engineer, and revised by the author, with numerous illustrations. 1 vol. 8vo. Cloth, $4.

A TREATISE ON THE RICHARDS STEAM-ENGINE INDICATOR, with directions for its use. By CHARLES T. PORTER. Revised, with notes and large additions as developed by American Practice, with an Appendix containing useful formulæ and rules for Engineers. By F. W. BACON, M. E., member of the American Society of Civil Engineers. 12mo. Illustrated. Cloth. $1

A COMPENDIOUS MANUAL OF QUALITATIVE CHEMICAL ANALYSIS. By CHAS. W. ELIOT, and FRANK H. STORER, Profs. of Chemistry in the Massachusetts Institute of Technology. 12mo. Illustrated. Clo. $1.50.

INVESTIGATIONS OF FORMULAS, for the Strength of the Iron Parts of Steam Machinery. By J. D. VAN BUREN, Jr., C. E. 1 vol. 8vo. Illustrated. Cloth. $2.

THE MECHANIC'S AND STUDENT'S GUIDE in the Designing and Construction of General Machine Gearing, as Eccentrics, Screws, Toothed Wheels, etc., and the Drawing of Rectilineal and Curved Surfaces; with Practical Rules and Details. Edited by FRANCIS HERBERT JOYNSON. Illustrated with 18 folded plates. 8vo. Cloth. $2.00.

"The aim of this work is to be a guide to mechanics in the designing and construction of general machine-gearing. This design it well fulfils, being plainly and sensibly written, and profusely illustrated."—*Sunday Times.*

FREE-HAND DRAWING: a Guide to Ornamental, Figure, and Landscape Drawing. By an Art Student. 18mo. Cloth. 75 cents.

THE EARTH'S CRUST: a Handy Outline of Geology. By DAVID PAGE. Fourth edition. 18mo. Cloth. 75 cents.

"Such a work as this was much wanted—a work giving in clear and intelligible outline the leading facts of the science, without amplification or irksome details. It is admirable in arrangement, and clear and easy, and, at the same time, forcible in style. It will lead, we hope, to the introduction of Geology into many schools that have neither time nor room for the study of large treatises."—*The Museum.*

HISTORY AND PROGRESS OF THE ELECTRIC TELEGRAPH, with Descriptions of some of the Apparatus. By ROBERT SABINE, C. E. Second edition, with additions. 12mo. Cloth. $1.75.

IRON TRUSS BRIDGES FOR RAILROADS. The Method of Calculating Strains in Trusses, with a careful comparison of the most prominent Trusses, in reference to economy in combination, etc., etc. By Brevet Colonel WILLIAM E. MERRILL, U. S. A., Major Corps of Engineers. With illustrations. 4to. Cloth. $5.00.

USEFUL INFORMATION FOR RAILWAY MEN. Compiled by W. G. HAMILTON, Engineer. Second edition, revised and enlarged. 570 pages. Pocket form. Morocco, gilt. $2.00.

REPORT ON MACHINERY AND PROCESSES OF THE INDUSTRIAL ARTS AND APPARATUS, OF THE EXACT SCIENCES. By F. A. P. BARNARD, LL. D.—Paris Universal Exposition, 1867. 1 vol., 8vo. Cloth. $5.00.

THE METALS USED IN CONSTRUCTION: Iron, Steel, Bessemer Metal, etc., etc. By FRANCIS HERBERT JOYNSON. Illustrated. 12mo. Cloth. 75 cents.

"In the interests of practical science, we are bound to notice this work; and to those who wish further information, we should say, buy it; and the outlay, we honestly believe, will be considered well spent."—*Scientific Review.*

DICTIONARY OF MANUFACTURES, MINING, MACHINERY, AND THE INDUSTRIAL ARTS. By GEORGE DODD 12mo Cloth. $2.00.

SUBMARINE BLASTING in Boston Harbor, Massachusetts—Removal of Tower and Corwin Rocks. By JOHN G. FOSTER, U. S. A. Illustrated with 7 plates. 4to. Cloth. $3.50.

KIRKWOOD. Report on the Filtration of River Waters, for the supply of Cities, as practised in Europe, made to the Board of Water Commissioners of the City of St. Louis. By JAMES P. KIRKWOOD. Illustrated by 30 double-plate engravings. 4to. Cloth. $15.00.

LECTURE NOTES ON PHYSICS. By ALFRED M. MAYER, Ph. D., Professor of Physics in the Lehigh University. 1 vol. 8vo. Cloth. $2.

THE PLANE TABLE, and its Uses in Topographical Surveying. From the Papers of the U. S. Coast Survey. 8vo. Cloth. $2.

FRENCH'S GRAMMAR. Part of a Course on Language. Prepared for Instruction in the U. S. Corps of Cadets. By Rev. J. W. FRENCH, D. D., Professor of Ethics and English Studies in the United States Military Academy, West Point. 1 vol. 12mo. Cloth. $2.50.

AGNEL. Elementary Tabular System of Instruction in French. Devised and arranged in practical form for the use of the Cadets of the U. S. Military Academy. By H. R. AGNEL, Professor of French. 1 vol. 8vo. Cloth, flexible. $3.50.

WILLIAMSON. Practical Tables in Meteorology and Hypsometry, in connection with the use of the Barometer. By Col. R. S. Williamson, U. S. A. 1 vol. 4to, flexible cloth. $2.50.

CULLEY. A Hand-Book of Practical Telegraphy. By R. S. CULLEY. Engineer to the Electric and International Telegraph Company. Fourth edition, revised and enlarged. 8vo. Illustrated. Cloth. $5.

POPE. Modern Practice of the Electric Telegraph. For Electricians and Operators. By FRANK L. POPE. Illustrated. 8vo. Cloth. $1.50.

GOUGE. New System of Ventilation, which has been thoroughly tested under the patronage of many distinguished persons. By HENRY A. GOUGE. Third edition, enlarged. With many illustrations. 8vo. Cloth. $2.

PLATTNER'S BLOW-PIPE ANALYSIS. A Complete Guide to Qualitative and Quantitative Examinations with the Blow-Pipe. Revised and enlarged by Prof. RICHTER, Freiberg. Translated from the latest German edition by HENRY B. CORNWALL, A. M., E. M. *In Press.*

www.ingramcontent.com/pod-product-compliance
Lightning Source LLC
LaVergne TN
LVHW021357110826
845150LV00007B/1687

* 9 7 8 1 4 2 5 5 1 3 7 6 4 *